Optimization of Pre-chamber Ignition Systems in Small Displacement Spark Ignition Engines to Increase Engine Efficiency

Zur Erlangung des akademischen Grades eines
Doktors der Ingenieurwissenschaften (Dr.-Ing.)

von der KIT-Fakultät für Maschinenbau des
Karlsruher Instituts für Technologie (KIT)

angenommene
Dissertation
von

Dipl.-Ing. Nicolas Johannes Karl Wippermann
aus Karlsruhe

Tag der mündlichen Prüfung: 29.11.2022
Hauptreferent: Prof. Dr. sc. techn. Thomas Koch
Korreferent: Prof. Dr.-Ing. Hermann Rottengruber

Forschungsberichte aus dem
Institut für Kolbenmaschinen
Karlsruher Institut für Technologie (KIT)
Hrsg.: Prof. Dr. sc. techn. Thomas Koch

Bibliografische Information der Deutschen Nationalbibliothek

Die Deutsche Nationalbibliothek verzeichnet diese Publikation in der
Deutschen Nationalbibliografie; detaillierte bibliografische Daten sind
im Internet über http://dnb.d-nb.de abrufbar.

ISBN 978-3-8325-5713-3
ISSN 1615-2980

Logos Verlag Berlin GmbH
Georg-Knorr-Str. 4, Geb. 10,
12681 Berlin
Tel.: +49 030 42 85 10 90
Fax: +49 030 42 85 10 92
http://www.logos-verlag.de

Vorwort des Herausgebers

Die Komplexität des verbrennungsmotorischen Antriebes ist seit über 100 Jahren Antrieb für kontinuierliche Aktivitäten im Bereich der Grundlagenforschung sowie der anwendungsorientierten Entwicklung. Die Kombination eines instationären, thermodynamischen Prozesses mit einem chemisch reaktiven und hochturbulenten Gemisch, welches in intensiver Wechselwirkung mit einer Mehrphasenströmung steht, stellt den technologisch anspruchsvollsten Anwendungsfall dar. Gleichzeitig ist das Produkt des Verbrennungsmotors aufgrund seiner vielseitigen Einsetzbarkeit und zahlreicher Produktvorteile für sehr viele Anwendungen annähernd konkurrenzlos. Nun steht der Verbrennungsmotor insbesondere aufgrund der Abgasemissionen im Blickpunkt des öffentlichen Interesses. Vor diesem Hintergrund ist eine weitere und kontinuierliche Verbesserung der Produkteigenschaften des Verbrennungsmotors unabdingbar. Am Institut für Kolbenmaschinen am Karlsruher Institut für Technologie wird deshalb intensiv an der Weiterentwicklung des Verbrennungsmotors geforscht. Übergeordnetes Ziel dieser Forschungsaktivitäten ist die Konzentration auf drei Entwicklungsschwerpunkte. Zum einen ist die weitere Reduzierung der Emissionen des Verbrennungsmotors, die bereits im Verlauf der letzten beiden Dekaden um circa zwei Größenordnungen reduziert werden konnten aufzuführen. Zum zweiten ist die langfristige Umstellung der Kraftstoffe auf eine nachhaltige Basis Ziel der verbrennungsmotorischen Forschungsaktivitäten. Diese Aktivitäten fokussieren gleichzeitig auf eine weitere Wirkungsgradsteigerung des Verbrennungsmotors. Der dritte Entwicklungsschwerpunkt zielt auf eine Systemverbesserung. Motivation ist beispielsweise eine Kostenreduzierung, Systemvereinfachung oder Robustheitssteigerung von technischen Lösungen. Bei den meisten Fragestellungen wird aus dem Dreiklang aus Grundlagenexperiment, Prüfstandversuch und Simulation eine technische Lösung erarbeitet. Die Arbeit an diesen Entwicklungsschwerpunkten bestimmt die Forschungs- und Entwicklungsaktivitäten des Instituts. Hierbei ist eine gesunde Mischung aus grundlagenorientierter Forschung und anwendungsorientierter Entwicklungsarbeit der Schlüssel für ein erfolgreiches Wirken. In nationalen als auch internationalen Vorhaben sind wir bestrebt, einen wissenschaftlich wertvollen Beitrag zur erfolgreichen Weiterentwicklung des Verbrennungsmotors beizusteuern. Sowohl Industriekooperationen als auch öffentlich geförderte Forschungsaktivitäten sind hierbei die Grundlage guter universitärer Forschung. Zur Diskussion der erarbeiteten Ergebnisse und Erkenntnisse dient diese Schriftenreihe, in der die Dissertationen des Instituts für Kolbenmaschinen verfasst sind. In dieser Sammlung sind somit die wesentlichen Ausarbeitungen des Instituts niedergeschrieben. Natürlich werden darüber hinaus auch Publikationen auf Konferenzen und in Fachzeitschriften veröffentlicht. Präsenz in der Fachwelt erarbeiten wir uns zudem durch die Einreichung von Erfindungsmeldungen und dem damit verknüpften Streben nach Patenten. Diese Aktivitäten sind jedoch erst das Resultat von vorgelagerter und erfolgreicher Grundlagenforschung. Jeder Doktorand am Institut beschäftigt sich mit Fragestellungen von ausgeprägter gesellschaftlicher Relevanz. Insbesondere Nachhaltigkeit und Umweltschutz als Triebfedern des ingenieurwissenschaftlichen Handelns sind die Motivation unserer Aktivität. Gleichzeitig kann er nach Beendigung seiner Promotion mit einer sehr guten Ausbildung in der Industrie oder Forschungslandschaft wichtige Beiträge leisten. In diesem Exemplar der Schriftenreihe berichtet Nicolas Wippermann über die Optimierung von Ottomotoren bei Verwendung einer Vorkammerzündung. Die definierten Haupteinflussgrößen Temperatur, Druck sowie Bewegung und Zusammensetzung des Gemischs innerhalb der Vorkammer stehen hierbei im Fokus der Untersuchungen. Während Druck und Temperatur durch Zylinderdrucksensoren und Thermoelemente relativ einfach zu messen sind wurde für die Bestimmung von Kohlenwasserstoffen innerhalb der Vorkammer auf optische Messtechnik zurückgegriffen. Mit Hilfe der Absorptionsmessung von Infrarotlicht konnten verbrannter und unverbrannter Kraftstoff innerhalb der Vorkammer kurbelwinkelgenau während des Arbeitsspiels aufgezeichnet werden. Die Bestimmung des Verbrennungsluftverhältnisses innerhalb der Vorkammer zum Zündzeitpunkt lässt Rückschlüsse auf die Kraftstoffversorgung durch unterschiedliche Einspritz-strategien zu. Zur Bestimmung der Strömung innerhalb der Vorkammer wurde ein Prüfstand konstruiert, der die Auslenkung des Zündfunkens mit Hilfe einer Hochgeschwindigkeitskamera aufnimmt. Durch unterschiedliche Bohrbilder von Kappen und Vorkammergeometrien kann hierfür die Strömung am Funken optimiert werden. In der vorliegenden Arbeit wird außerdem auf den Einfluss auf die Motorreglung, potentielle Haltbarkeitsprobleme des Zylinderkopfes und der Vorkammer selbst eingegangen. Die Arbeit endet mit einer Orientierungshilfe zur Entwicklung eines Vorkammerzündsystems. Hier wird in fünf Schritten der Entwicklungsablauf dargestellt in welchem die einzelnen Entwicklungsschritte der Arbeit aufgegriffen werden.

Karlsruhe im Mai 2023 Prof. Dr. sc.-techn. Thomas Koch

Vorwort des Autors

Die vorliegende Arbeit entstand während meiner Anstellung als Entwicklungsingenieur im Formel 1 Programm bei Alpine Racing in Viry-Châtillon in einem Forschungsprojekt mit dem Institut für Kolbenmaschinen des Karlsruher Institut für Technologie, bei dem ich als Gastwissenschaftler tätig war. Mein besonderer Dank gilt Prof. Dr. sc. techn. Thomas Koch für seine Unterstützung und die Betreuung meiner Arbeit. Bei Prof. Dr.-Ing. Hermann Rottengruber bedanke ich mich für das Interesse an meiner Arbeit und die Übernahme des Korreferats. Frau Prof. Dr. Dr.-Ing. Dr. h. c. Jivka Ovtcharova danke ich für die Übernahme des Vorsitzes bei der mündlichen Prüfung.

Je remercie chaleureusement Rémi Taffin, Vincent Hubert et Jean-Philippe Aignan de m'avoir accordé leur confiance dans ce projet et m'avoir permis de l'effectuer en parallèle de mon rôle d'ingénieur de développement. Je remercie l'équipe Culasse, Distribution et Allumage d'Alpine pour son soutien et sa compréhension. Je remercie Hubert De-Sousa pour sa collaboration constructive et ses performances exceptionnelles lors du développement de l'allumage chez Alpine. Cette expérience marquera à jamais ma carrière professionnelle et ma vie personnelle. Je remercie Emmanuel Labussière pour ses nombreuses inspirations et sa contribution à mon apprentissage de la langue française. Je remercie Julie Colling de sa compréhension pendant la thèse et d'avoir été une cheffe parfaite. C'est une tristesse profonde que tu sois partie si tôt. Je remercie également Alexandre Borie pour ses conversations instructives et ses idées géniales, Alexandre Hebert de sa confiance au fil des ans et Frédéric Castex de son soutien, particulièrement dans la phase finale du projet. Je remercie Franck Dubois de la poursuite du développement et la bonne coopération que nous entretenons. Enfin, je remercie également tous les mécaniciens et ingénieurs qui m'ont aidé à réaliser ce travail. Mes années passées en France à vos côtés resteront à jamais un souvenir heureux, gravé dans ma mémoire. Merci infiniment pour ce moment extraordinaire.

Am Institut möchte ich besonders den Herren Dr. Olaf Toedter und Tobias Michler danken, mit denen ich meinen Horizont für das Thema erweitern konnte und mit denen ich gleichgesinnte Zündungsverrückte gefunden habe. Ich freue mich auf zukünftige spannende Jahre. Am IFKM möchte ich außerdem den Herren Alexander Heinz, Johannes Dornhöfer und Thomas Wheying, Moritz Grüniger und Jan Reimer für die freundliche Aufnahme am IFKM und die Hilfe vor Ort in Karlsruhe danken. Außerdem danke ich Herrn Michael Busch für den Support mit der Internetverbindung über den Rhein und Julia Reichelt für die organisatorische Unterstützung am KIT.

Mein herzlicher Dank gebührt der Multitorch GmbH, die mich während des Projekts mit Versuchsteilen und Rat unterstützt hat. Herrn Dr. Steffen Kuhnert danke ich besonders für die vielen Diskussionen und die andauernde Unterstützung bei dem Projekt.

Der Firma LaVision und besonders Dr. Olaf Thiele danke ich für die Unterstützung mit der beeindruckenden Messtechnik und den technischen Austausch bei der Auswertung.

Ganz besonders danke ich meiner Familie. Meinem Vater Klaus für das Wecken und Fördern meiner Begeisterung für Technik und für die vielen Gespräche und das Interesse an meiner Arbeit. Meiner Mutter Angela danke ich für meine Offenheit für Sprachen und dadurch das Vertrauen in mich nach Frankreich zu ziehen. Meinen Geschwistern Robert, Christina und Magdalena für den tollen Zusammenhalt seit unserer Kindheit und die aufmunternden Worte wenn sie nötig waren. Ich bin sehr stolz auf jeden von uns. Meinen Schwagern und meiner Schwägerin und allen sieben Neffen und Nichten danke ich für die schöne Zerstreuung in der Freizeit und den Familienzusammenhalt.

Mein größter Dank gilt meiner Frau Sabrina, die mich während den unzähligen Stunden der Doktorarbeit unterstützt und motiviert hat. Die mir zusätzlich zu der Arbeit in Frankreich die ermöglicht hat den Wunsch einer Dissertation neben dem Beruf zu verwirklichen. Unserer wundervollen Tochter Amélie danke ich für die letzte notwendige Motivation zum Fertigstellen der Arbeit. Ich liebe euch und freue mich auf unsere Zukunft.

Contents

Kurzfassung

Ziel der vorliegenden Forschungsarbeit ist die Optimierung des fremdgezündeten Motorbetriebs durch den Einsatz einer Vorkammerzündung. Dabei wird der Hauptfokus neben der Steigerung des thermischen Wirkungsgrads, auf die Verkürzung der Brenndauer und die Reduktion von zyklischen Schwankungen gelegt. Als Haupteinflüsse auf das Verhalten der Vorkammer wurden definiert: Gemischzusammensetzung, Temperatur, Druck und Luftbewegung innerhalb der Vorkammer im Zusammenspiel mit dem Design der Vorkammerzündkerze. Nach Vorbild des üblichen Entwicklungsablaufs wurden Simulationen von unterschiedlichen Vorkammern durchgeführt und Varianten an Prüfständen in Karlsruhe und Viry-Châtillon untersucht. Die Prüflinge unterschieden sich hierbei in den verwendeten Materialien aber teilweise auch stark in ihrer Konstruktion. Bei der Analyse wurden auch unterschiedliche Lastbereiche und Umgebungsbedingungen für den Fahrzeugeinsatz berücksichtigt. Die Ergebnisse der Untersuchungen legen nahe, dass die einströmende Luftbewegung entscheidend für die Leistung der Vorkammer ist. Um deren Einfluss auf die Funkenbildung zu quantifizieren, wurde im Rahmen der Forschung am IFKM eine Funkenauslenkprüfkammer für Vorkammerzündkerzen entwickelt. Während der Parametervariation wurde die höchste Funkenauslenkung durch einen kleinen Innendurchmesser der Vorkammer und einen erhöhten Lochversatz erreicht. Im Prüfstandbetrieb erwies sich die Druckmessung innerhalb der Vorkammer zusätzlich zu der im Brennraum als wertvolles Instrument für ein vertieftes Verständnis über den Gasaustausch zwischen beiden Volumina. Das Wissen über die parallele Druckentwicklung in der Vorkammer und dem Hauptbrennraum hilft, das richtige Verhältnis zwischen Vorkammervolumen und Luft-Kraftstoff-Verhältnis für einen optimalen Betriebspunkt zu finden. Für die Auslegung der Vorkammerkomponenten können so außerdem mechanische Grenzen bestimmt und die Ursachen von Motorstörungen identifizieren werden. Durch die Kenntnis der Temperatur der Vorkammer können potentielle Entstehungsorte für Oberflächenentzündungen bestimmt werden. Die Oberflächentemperatur der Mittelelektrode wurde hierbei als Hauptursache für Vorentflammungen identifiziert. Die Auswirkungen des Luft-Kraftstoff-Verhältnisses wurde bei optimalem Zündwinkel als Haupttreiber für die Zündkerzentemperatur ermittelt. Die Analyse der Gemischaufbereitung innerhalb der Vorkammer erfolgte mithilfe eines Infrarot-Sensors. Aufgrund der Absorption des Lichts durch Kohlenwasserstoff- und Kohlendioxidmoleküle ist die Identifizierung von unverbranntem und verbranntem Kraftstoff innerhalb der Vorkammer möglich. Eine starke Abhängigkeit des Kraftstoffgehalts in der Vorkammer vom Einspritzbeginn wurde bestätigt. Ein überraschendes Ergebnis ist, dass in der Vorkammer am Ende des Arbeitszyklus keine vollständige Kohlendoxid-Füllung gemessen wurde. Die Erkenntnisse aus den Untersuchungen der vorliegenden Arbeit wurden in einem Leitfaden zur Entwicklung einer Vorkammerzündung zusammengefasst.

1 Introduction

The development of internal combustion engines (ICE) is an ongoing process over the last century. The increase in interest about the global warming, together with a higher acceptance of the population to change towards environmental protection, challenge the engineers and companies that work for the optimization of thermal engines. Since the beginning of the mass production of passenger cars and their powertrains, the internal combustion engine using a diesel or otto process was the undisputed choice. During the last decades, stricter emission allegations and carbon dioxide (CO_2) targets pushed engine developers towards new solutions, for example the direct fuel injection for mass-produced cars in 1996 by Mitsubishi motors [87] or the begin of the downsizing of engines in the 2000s [42]. Due to the efforts to reduce the fuel consumption of the vehicles, the increase of the stock of cars does not augment the overall CO_2 emissions of cars as shown in Figure 1.1. The increase of the ICE efficiency in combination with the treatment of the combustion emissions are therefore key for the engine development.

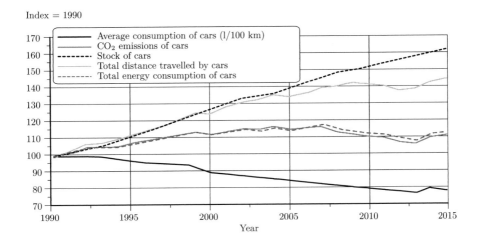

Figure 1.1: Fuel efficiency and fuel consumption trends for private cars in the EU-28 in the period 1990 to 2015. Adapted from the European Environment Agency [37]

Since the beginning of the 20th century, a new trend is challenging the ascendance of the ICE. Electric vehicles use local emission free powertrains, where the ICE is replaced by an electrical motor. Their electrical energy source consists of either an accumulator or a fuel cell.

Besides of a total replacement of the combustion engine, the hybridization combines the ICE with electrical components to increase the overall efficiency of the powertrain. Different automotive manufacturers such as Renault are using their motorsport engagement to test and develop these complex powertrains. As motorsport is used for marketing of road cars, they are strongly dependent on their image and trends in the targeted population. An example of the upcoming connection between motorsport and the environment is the work of Schwarz, who investigated different propulsion concepts in regard to their CO_2 impact in motorsports [104]. Rule maker of motorsport series, like the Fédération Internationale de l'Automobile (FIA), are creating the regulations to give the necessary freedom for the car companies for developing the future powertrains. Therefore the rules are build around a wish of the car companies to develop more road car relevant propulsion systems and to build up further competence to increase the efficiency of the ICE.

While in the past the power of high level motorsport engines was limited either by an air restrictor or by their displacement and a maximal engine speed together with a defined maximal boost pressure or even an interdiction of supercharging [38], the regulations changed to a more engine efficiency related approach. Since 2014 the FIA Formula 1 (F1) rules limit the fuel mass flow of the engine to 100 kg/h, thus the engine power is principally defined by the engine efficiency for a given fuel and oil consumption. The same principal is also used or adapted for several other race series such as the World Endurance Championship, Deutsche Tourenwagen Masters and Super GT series [33, 39, 58].

In F1 the regulation change results in powertrains with over 50 % efficiency. The ICE in the F1 cars reaches values of about 45 % thermal efficiency [127]. Key for the performance of the ICE is an optimized overstochiometric combustion. To increase the power output of the lean operation points (OPs), different strategies are developed. Mandatory for a competitive efficiency level seems to be the introduction of a pre-chamber ignition system [110]. The gain by the pre-chamber is not only the increase of the combustion efficiency but it permits also to unblock operation fields that are not accessible with a standard spark plug[1]. The first F1 team that introduced the innovative ignition was Mercedes giving them a significant development and performance advantage. After the information about the introduction of a pre-chamber combustion spread, their potential use was investigated in other motorsport classes [25].

However, the spark ignition (SI) inside a pre-chamber is not a recent invention but subject of various research projects. It was used in road cars in the 1970s by Honda for their Compound Vortex Controlled Combustion [123] and is a common ignition system for larger gas engines. The interest in the pre-chamber ignition is rising over the last years. This is also documented by the number of pre-chamber relevant topics on the International Conference on Ignition Systems for Gasoline Engines in Berlin, where the number increased from one in 2016 to nine in 2018 [46, 47]. The pre-chambers that are used in SI engines can be divided in active and passive systems. An active pre-chamber has a dedicated fuel supply to enrich the mixture around the electrodes, where the flame kernel grows. The richer mixture is beneficial for a fast and stable combustion [48]. As the air-fuel ratio in the pre-chamber is independent of the mixture in the main combustion chamber, the global mixture can be significantly diluted.

---

[1]For example higher compression ratios or extreme mixture movements inside the combustion chamber

Passive pre-chamber systems do not have a dedicated fuel supply at their disposal. The fuel that is used for the pre-chamber combustion is injected outside its volume into the inlet ports or the main combustion chamber (MCC), therefore the lean limit occurs earlier than with an active system [105].

As FIA F1 regulations do not allow an active system, this work focuses on the development of passive pre-chambers. However, most results can be transferred and investigation methods can be adapted for an active system.

Common for both pre-chamber types is the use of a relatively small amount of fuel energy for a premature combustion in the separate volume. The defined values of Gussak et al. in the 1970s give an orientation about the design of a pre-chamber with a size of 2 % - 3 % of the compression volume together with an orifice surface to volume ratio of $0.03\,cm^{-1}$ - $0.04\,cm^{-1}$ and length to diameter ratio of 1:2 [48, 49]. The pre-chamber is connected with holes to the MCC, the pressure rise in the pre-chamber creates hot gas jets that protrude into the MCC and ignite the mixture at multiple points. The result is a faster combustion with less cyclic variations.

Although the use of pre-chamber ignition systems in SI engines is a well known technique to increase efficiency of larger gas engines, its introduction in motorsport shows the potential to reduce fuel consumption and emissions for more road relevant engines. Due to the hype that was caused by the F1 pre-chambers in the years 2014-2018, several engine manufacturers and suppliers focus on the development of pre-chamber ignition systems. The development principals of the ignition system can be inspired by the existing knowledge base of the larger engines and need to be validated or adapted for engines with smaller displacements. New aspects that occur during this transfer of technology are focus of the research in this work.

The data that contribute to the results in this work is collected on test benches at Renault Sport Racing (RSR) in Viry-Châtillon (France) and at the Institute of Internal Combustion Engines (IFKM) at the Karlsruhe Institute of Technology (KIT) in Karlsruhe (Germany). The investigation focus on the achievement of new analysis methods and results to optimize the performance of the pre-chamber ignition systems.

The principal drivers that impact a pre-chamber ignition in an SI engine and their origin are displayed in Figure 1.2. The two volumes of MCC and pre-chamber exchange gas via orifices that connect them both.

The performance and reliability of the pre-chamber mainly depend on the mixture composition (air-fuel equivalence ratio (λ) and CO_2), the state of the gas (pressure and temperature), the composition of the used fuel and air, but also the air movement into and out of the pre-chamber. The latter is controlled by the orifice design such as number of holes, hole diameter and hole orientation. The gas movement inside the pre-chamber has a significant impact on the combustion inside the pre-chamber. The right positioning of the the hot gas jets, which ignite the mixture in the MCC, is key for the performance of the engine.

The pressure in the MCC creates the gas flow into the pre-chamber and sets the level on which the pre-chamber combustion pressure is added. Pressure and temperature that are imposed by the combustion in the MCC define the reliability requirements of the pre-chamber and are therefore essential design input parameters.

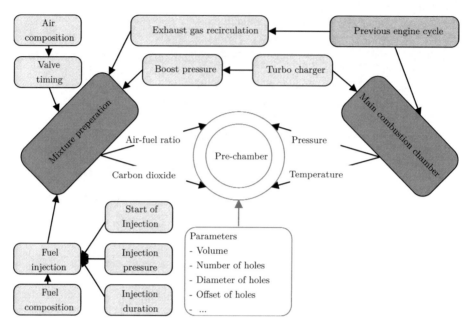

Figure 1.2: Principle impacts on the pre-chamber in a spark ignition engine. The schema shows the connection between the different influences that are investigated in the present work

The mixture preparation is driven by the injection of fuel and the scavenging during the gas exchange of the MCC and pre-chamber. For active pre-chambers the internal air-fuel ratio can be directly controlled, passive pre-chambers depend on the mixture around their connection holes in the MCC.

The fuel that contributes to the combustion in the pre-chamber is coming from an injection with the common parameters such as injection duration, start of injection or fuel pressure and forms the combustible mixture together with the air in the cylinder. CO_2, which remains or enters the pre-chamber is left over from the previous engine cycle or from an external exhaust gas recirculation. The turbo charger creates the boost pressure, which increases the density of the fresh air but also increases the burned residuals in MCC, due to the increased exhaust pressure of a turbo engine.

In this work the different main drivers of the scheme in Figure 1.2, which have an impact on the pre-chamber combustion, are investigated. In combination with an analysis of major pre-chamber design parameters such as materials and orifice designs, knowledge is created for the use of the ignition system in a car application.

To determine the local air-fuel ratio inside the pre-chamber, a pre-chamber spark plug was developed using an infrared (IR) sensor to measure the fuel density, due to the absorption of light by the hydrocarbons. The same principal is used for the local concentration of CO_2 inside the pre-chamber.

The pressure measurement inside the pre-chamber volume is done via an integration of piezoelectric pressure sensors. The knowledge of the pressure in MCC and pre-chamber helps to understand the gas exchange between both volumes.

The temperature of the pre-chamber is measured with thermocouples.

As the orifice design is crucial for the functioning of the pre-chamber ignition, a test chamber was developed to quantify the caused internal movement inside the pre-chamber by the inflowing gas. Focus is hereby the deflection of the plasma between the spark plug electrodes. The ignition of the mixture in the MCC by the hot gas jets is investigated via multiple experiments with different pre-chamber designs both at RSR and at the IFKM. The development is supported by common simulation techniques that are correlated with the data from the measurement. The use of investigation methods such a high speed camera recording to visualize the jets and the spark are giving additional insights. The reader is given assistance for the dimensioning and the mechanical design of the passive pre-chamber ignition system.

As the well operating pre-chamber significantly changes the combustion in the MCC, the control mechanism of the engine needs to be adapted. The challenges that occur hereby, but also the benefits of the jet ignition are part of the present work.

2 Fundamentals of engine efficiency and pre-chamber systems of spark ignition engines

2.1 Combustion efficiency principals and their impact on pre-chamber designs

One principle target of engine development is the increase of the engine efficiency. In this chapter the fundamentals of possible levers and their impact on the pre-chamber design are described.

The most common way to characterize the efficiency of an internal combustion engine is the comparison of its power output with the available fuel power. The fuel efficiency (η_f) is defined as the fraction of the delivered work per cycle (W_e) compared to the maximal available work from its fuel [92]. The fuel energy is determined by the fuel mass (m_f) and its heating value (Q_{HV}). In an ICE the lower heating value at constant pressure is used, as the water in the exhaust is always in vapor form and the heat of its vaporization is not recovered during the engine cycle [55].

$$\eta_f = \frac{W_e}{m_f \cdot Q_{HV}} = \eta_i \cdot \eta_m \tag{2.1}$$

The work in Equation 2.1 can be measured on the crank shaft and includes all engine losses such as friction or pumping during the gas exchange. It is therefore the product of the mechanical efficiency (η_m) and the indicated efficiency (η_i). The mechanical efficiency can be increased by reducing the friction of the engine. For a detailed look at combustion and gas exchange, the indicated work (W_i) excludes these mechanical aspects. Therefore, the indicated efficiency is more relevant for the optimization of the combustion process, as focused in this work. The indicated work is calculated via a cylinder pressure measurement and the cylinder volume. For four-stroke engines, the indicated work can be divided into a high pressure and a low pressure loop. The high pressure loop consists of compression and expansion stroke, the low pressure loop of intake and exhaust stroke. The heat that is supplied in an ICE contributes in the high pressure cycle and creates the positive work of the engine. The low pressure loop is responsible for the gas exchange. In a naturally aspirated engine the pressure level during the exhaust is higher than during intake stroke, the resulting counter clockwise pressure loop results in a negative work. Charged engines with a forced induction can have a positive low pressure loop, due to their boost pressure.

An additional benefit of higher air-fuel ratios is the increased pressure level during the time of the open intake valve.

2.1.1 Efficiency reflection in closed engine cycles

To visualize the drivers that increase the efficiency of the fuel transformation in the ICE, it is helpful to reduce the complexity of the complete engine. The open four stroke cycle is therefore reduced to an ideal closed engine cycle. The gas exchange is substituted by a heat removal (q_{rem}) and following simplifications are made [92]:

- Combustion is simplified by the use of a constant volume, constant pressure or limited pressure cycle. The cycles are hereby named after the moment when heat is added during the cycle.

- Heat transfer and friction is not considered (isentropic compression and expansion).

- Gas exchange does not include throttle losses and all gas is exchanged during the valve opening timings at the dead centers.

- The working gas in the engine is assumed to be an ideal gas with constant properties (κ, c_p, c_v, R) .

The maximal achievable thermal efficiency of an ideal heat engine is defined by the Carnot process. Carnot describes a reversible engine cycle in which heat is transferred from a hot reservoir to a colder one, converting some of the energy to mechanical work by an isolated piston [24]. The idea of his engine was the basis for Carnot's theorem in Equation 2.2. The efficiency of the the the engine ($\eta_{th,C}$) depends hereby on the level and difference between the maximal (T_{max}) and minimal Temperature (T_{min}) in the cycle.

$$\eta_{th,C} = 1 - \frac{T_{min}}{T_{max}} \tag{2.2}$$

For temperatures that occur in an ICE, Carnot's engine would result in an theoretical efficiency of 70 % [116]. Even if the thermodynamic cycle is not realizable, it shows the upper efficiency frontier for heat engines. Contrary to Carnot's theoretical engine, the thermodynamic process in an ICE is not reversible and not a closed system.

The highest realizable efficiency in an ICE can be achieved with a constant volume cycle. The transformation efficiency of work from available heat is hereby defined by the compression ratio (ε) and the heat capacity ratio (κ), which connects the heat capacity at constant pressure (c_p) to the heat capacity at constant volume (c_v). Formula 2.3 shows the importance of the compression ratio for an efficient SI engine.

$$\eta_{th,v} = 1 - \frac{1}{\varepsilon^{\kappa-1}} \tag{2.3}$$

In the constant volume cycle the heat is added during the moment with the smallest volume. In an ICE this would mean an infinitive fast combustion at top dead center

(TDC) with a resulting very high cylinder pressure. Due to the needed time for the real combustion and also to limit the maximum pressure for mechanical reasons, the constant volume cycle can only be approximated. For former SI engines, the constant volume cycle is often chosen as its reference cycle, it is therefore also called the otto cycle, named after its inventor Nicolaus August Otto. In diesel engines the constant pressure cycle is used, due to the combustion of the fuel that is injected after the compression and the high maximal pressure that needs to be limited[1]. The cycle is defined by a heat addition during the volume increase after TDC. The heat compensates the pressure loss due to the down moving piston resulting in a constant pressure. Modern SI engines reach comparable pressure levels as diesel engines, thus an extension of the constant volume cycle is necessary. The result is the more realistic model, called mixed cycle or Seiliger cycle. Engines with the same compression ratio are always less efficient with a pressure limitation such as defined by the constant pressure cycle. In addition to ε and κ, the efficiency equation ($\eta_{th,m}$) is extended by the pressure at bottom dead center (BDC) (p_1) and TDC (p_3), the temperature at BDC (T_1), c_p and the supplied heat (q_{suppl}).

$$\eta_{th,m} = 1 - \frac{\left[\frac{q_{suppl}}{c_p \cdot T_1} - \frac{1}{\kappa\varepsilon}\left(\frac{p_3}{p_1} - \varepsilon^\kappa\right) + \frac{p_3}{p_1\varepsilon}\right]^\kappa \left(\frac{p_1}{p_3}\right)^{\kappa-1} - 1}{\kappa \cdot \frac{q_{suppl}}{c_p \cdot T_1}} \tag{2.4}$$

Equation 2.4 shows that a high compression ratio ε, maximal pressure p_3, and a reduced heat supply (high air-fuel ratio) increase the efficiency [92].

The comparison of the different cycles in a pressure-volume (p-V) and a temperature-specific entropy (T-s) diagram is shown in Figure 2.1. Additional removed heat (q_{rem}) represents a reduction in efficiency, as less energy is transformed into work.

For the simplified calculation with constant heat capacities, the air-fuel ratio has no impact on the engine efficiency (Equation 2.3 and 2.4). However, in a real engine the air-fuel ratio is a major driver for the latter. Therefore the calculated efficiency (with Equation 2.4) is only approximately reached for very lean engine conditions. There are major errors for pressure and temperature compared to the detailed calculation [92]. To increase the accuracy of the calculation, the ideal engine after DIN 1940 was introduced. Hereby the air-fuel ratio and the actual properties of the working gas are taken into account. This includes the transformation of the ideal gas composition[2] and the temperature dependency of the gas properties. Figure 2.2 shows the impact on the heat capacity ratio (κ) for different temperatures, burned gas fractions (x_b) and air-fuel equivalent ratios (λ). The graph illustrates the interest of a high air-fuel ratio to increase κ and hereby the efficiency of the engine and gives an idea about the evolution of κ during the combustion in the engine cycle.

---

[1]The increased cylinder pressure is a result of the higher compression ratio of the diesel engine
[2]from unburned to burned gas due to the combustion

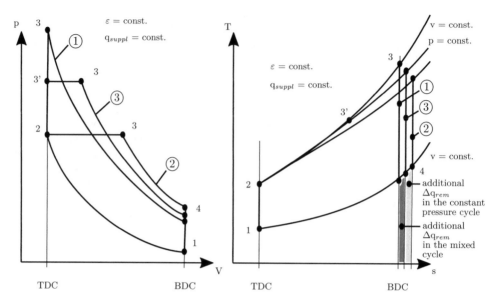

Comparison of the closed cycles,

(1) = constant volume cycle, (2) = constant pressure cycle, (3) = mixed cycle

Figure 2.1: Comparison of the different engine cycles in the p-V and T-s diagram adapted from Merker et al. [84]

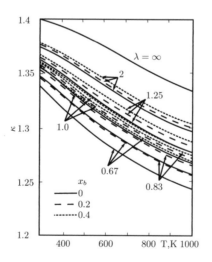

Figure 2.2: Heat capacity ratio (κ) as function of temperature (T), air-fuel equivalence ratio (λ) and burned gas fraction (x_b) adapted from Heywood [55].

2.1.2 Efficiency reflections in real engine combustion cycles

The simplified engine cycles in section 2.1.1 show the principals of an efficient ideal engine cycle. To visualize losses, which occur during the combustion in the engine, the different steps from a constant volume cycle to the real combustion are shown in Figure 2.3.

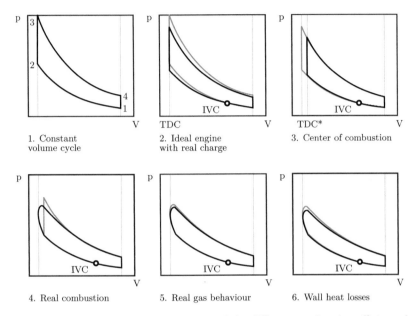

Figure 2.3: Step-by-step processes comparison of the different combustion efficiency losses in p-V diagrams adapted from Weberbauer et al. [125]

The investigation method that is used by Weberbauer et al. is applied for different types of engines, including diesel and controlled auto ignition (CAI) engines. Hereby, the main differences between the theoretical efficiency of an ideal combustion process (Graph 1 in Figure 2.3) and a combustion in a real engine are the following [125]:

- Not 100 % of the fuel energy is transformed into heat. Even though for lean conditions no carbon monoxide (CO) is formed, hydrocarbon (HC) emissions tend to increase. (Graph 2 in Figure 2.3)

- A constant volume cycle with reduced compression ratio is used instead of the mixed cycle to represent the targeted engine. (Graph 3 in Figure 2.3)

- The combustion in an engine is not indefinitely fast at TDC. The pressure increase in the cylinder by the shape and duration of the real combustion, adds further losses. (Graph 4 in Figure 2.3)

- The real gas behavior for air, burned gas residuals and fuel depends on pressure and temperature. (Graph 5 in Figure 2.3)

- Wall heat losses that are transferred through the combustion chamber walls during the expansion stroke. In the real engine a faster combustion results also in higher heat losses. (Graph 6 in Figure 2.3)

- Expansion and compression losses, which occur due to the exhaust valve opening before and intake valve closing after BDC.

Mechanical losses due to friction and pumping losses during the gas exchange do not directly contribute to the combustion efficiency and are not listed.

Figure 2.3 shows the different losses explained in p-V diagrams. As the area inside the clockwise process represents the work of the engine, each reduction stands for a decrease in efficiency. The closing of the intake valves is marked with IVC. The graphs show the losses and therefore the decrease in thermal efficiency. Weberbauer compares the losses of different engine types at 2000 rpm and 2 bar break mean effective pressure [125].

The results show the potentials and challenges of each combustion process in Figure 2.4. The diagram indicates the increase in heat losses of a fast CAI combustion compared to the direct fuel injection (DFI). The main difference, between manifold port injection (MPI) and DFI, is the working gas and the cooling effects of the vaporization of the fuel and during the gas exchange. The combustion in a diesel engine with direct injection (DDI) combines the positive effects of a fast combustion with less temperature losses due to a higher air-fuel ratio and hereby a lower temperature level in the combustion chamber.

The reflections of Section 2.1.1 in addition with the upstanding losses are giving the principal indication for objectives in the combustion development.

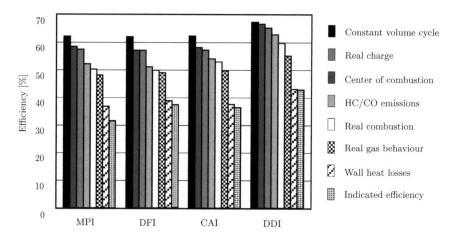

Figure 2.4: Comparison of efficiency respectively process losses of different engine types adapted from Weberbauer et al.[125]

This includes a combustion at high air-fuel ratios, a fast heat supply in an engine at a high compression ratio and no maximum pressure limitation. The mechanical challenges in the engine development are hereby imposed. The parts that define the combustion volume (e.g. cylinder head, piston, spark plug, valves, direct injector) and the parts that support

them (e.g. connection rod, crankcase, crankshaft) need to be designed to withstand a high maximum pressure respectively the resulting force. The reduction of thermal losses, due to different materials or coatings of the combustion chamber can also help to increase efficiency [63].
The combustion development should focus on the speed increase of the heat transformation from fuel to work at high air-fuel ratios.

2.1.3 Combustion challenges in charged lean burn engines

The desired fast energy transformation in an efficient ICE depends on its fast combustion by the moving flame trough the combustion chamber. The velocity of the flame is driven by its laminar and turbulent flame speed. The laminar flame speed is mainly dependent on conditions in the combustion chamber and the chemistry of the mixture. The turbulent flame speed adds the impact of the flow field in the combustion chamber to the flame propagation. The flow field is hereby imposed by micro and macro movements, such as tumble or swirl in the MCC.
Flamelet-models consider that the propagation of the turbulent flame front is an ensemble of multiple laminar flame fronts (Figure 2.5).

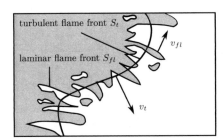

Figure 2.5: Flamelet model with turbulent and laminar flame velocities (v) and flame surface (S) adapted from R. Pischinger [92]

The mass transformation in the propagating flame can be described with the approach of Damköhler for the laminar to turbulent surface ratio [92] by Equation 2.5 . It is dependent on the density of the unburned gas ρ_u, the laminar flame speed v_{fl} and the surface of the turbulent flame front S_t. :

$$dm/dt = \rho_u \cdot v_t \cdot S_t \qquad (2.5)$$

The speed of the turbulent flame (v_t) that travels trough the combustion chamber is hereby the sum of the laminar velocity (v_{fl}) and the turbulent fluctuation velocity in the unburned mixture (v'):

$$v_t = v_{fl} + v' \qquad (2.6)$$

The laminar flame speed depends on transportation processes, namely the thermal conductivity ($a = \lambda/\rho c_p$) and the diffusion of radicals. Equation 2.7 shows that the laminar flame speed depends on one hand on the thermal conductivity (a) and on the other hand on the reaction time (τ_{ch}):

$$v_{fl} = \sqrt{a/\tau_{ch}} \tag{2.7}$$

At ambient conditions, the laminar flame speed for hydrocarbons-air mixtures is about 40 cm/s [92], it increases with temperature and decreases with pressure. The highest laminar flame speed occurs at a λ of around 0.9, for leaner or richer mixtures, the flame speed decreases. The laminar flame speed for different engine conditions was investigated by Teodosio et al. in a numerical approach [115]. The graphs in Figure 2.6 show the results of the experimental derived correlation [115].

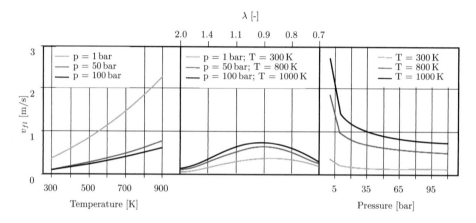

Figure 2.6: Laminar flame speed (v_{fl}) as a function of air-fuel equivalence ratio (λ), pressure (p) and temperature (T) adapted from Teodosio et al. [115]

Unfortunately lean burn ICEs combine different negative effects on the laminar flame speed:

- Low cylinder temperatures (e.g. due to a Miller cycle) to prevent knocking.

- High pressure in the combustion chamber, due to high boost pressure and high compression ratio

- Highly deluded mixture

Therefore the decrease of the laminar flame front velocity needs to be compensated by a higher turbulent fluctuation velocity (Equation 2.7). A higher fluctuation velocity can be achieved by adding turbulent kinetic energy (TKE) to the combustion and therefore increases the surface of the laminar flame front. TKE results out of a macro air movements in the combustion chamber (e.g. tumble or swirl or squish). The larger eddies in the chamber are broken down by the proximity of the piston and cylinder head to smaller eddies, of which each later increases the surface of the propagating flame. The increase of the turbulence in an ICE helps to decrease the combustion duration significantly [55]. For a given intake port design, a higher engine speed increases the turbulence in the engine, thus the combustion duration in degree crank angle (°CA) remains almost constant. However, a too intense turbulence can extinguish the early flame kernel [112]. Methods that decrease the cylinder temperature to prevent knocking (e.g. Miller cycle) may also have a strong negative impact on the turbulence field in the combustion chamber, as the inlet flow is reduced to a short duration with a piston position close to TDC, thus a tumble in the cylinder can not be developed and transformed into TKE.

Beside the flame velocity, the combustion chamber geometry has an impact on the flame propagation [92]. The origin of the flame is the electrode gap of the spark plug, its position together with the knowledge of the mixture and flow field is crucial for a symmetrical propagation through the combustion chamber to avoid knocking. High compression ratios with a resulting flat combustion chambers are also not optimal for the spherical flame propagation [122]. Quenching can occur close to the walls – especially with lean mixtures – and result in HC emissions, hence fuel energy which does not contribute to the engine's work output [92].

2.1.4 Cyclic variation impact on efficiency

The combustion in an ICE is not a constant process, but is changing after its initiation as the conditions in the combustion chamber vary from cycle to cycle. Origins for these variations can be divided in mixture composition, fluctuations of the cylinder charge or flow field and deviations of ignition parameters. In the engine development it is common to use the coefficient of variation (COV) of the indicated mean effective pressure (IMEP) to classify the variations. The standard deviation σ_{IMEP} is hereby divided by the IMEP as stated in Equation 2.8:

$$COV_{IMEP} = \frac{\sigma_{IMEP}}{IMEP} \cdot 100\,\% \tag{2.8}$$

Especially for higher air-fuel ratios or increased residual burned gas, the cyclic variation is a limiting parameter for OPs. As the combustion initiation by the spark plug in the combustion chamber is local, there is a strong dependency on the mixture conditions between the electrodes at ignition. Lean mixtures with their reduced fuel transformation speed are increasing these effects. With increased cyclic variations, the adjustment of the ignition timing also becomes more difficult. In a cycle with a richer mixture around the spark plug, the adjusted ignition timing might cause a knock event due to the faster combustion. Cycles with leaner local λ can cause misfires or a delayed combustion. When an engine is operated at its knock limit, the correct ignition timing is key for the performance and reliability. So it is desirable to reduce the COV_{IMEP} to a maximum for a given OP.

Besides the proper engine control, the reduction of the cyclic variations has an impact on the mechanical design of the engine. A smaller standard deviation of the cylinder pressure (p_{Cyl}) narrows down the operation range of the loaded parts. If the average maximum pressure is used for the design of the engine parts, high cylinder pressures due to cycles with a very fast combustion and the cycles with low pressure are equalized. Adding pressure curves, which represent two times the standard deviation of the cylinder pressure ($\pm 2\sigma_{p_{Cyl}}$) provides necessary information about a more realistic charge of the engine parts, as the envelop of the $\pm 2\sigma_{p_{Cyl}}$ pressure curves represent about 95.45 % of the occurring pressures[3].

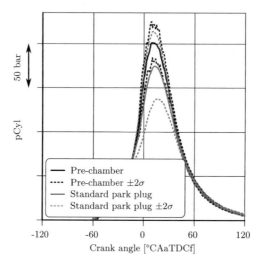

Figure 2.7: Comparison of the maximal expected cylinder pressure for a pre-chamber and a standard spark plug. The pressure envelope of the standard deviation is displayed together with the average pressure of 300 cycles

An example for the criteria is shown in Figure 2.7 where the pressure curves of OPs with a standard spark plug are compared to the pre-chamber ignition. Despite the higher average pressure of the jet ignition, the design limits of the engine parts are not increased as the $+2\sigma_{p_{Cyl}}$ remains the same for both ignition systems.
The reduction of the cyclic variations helps to raise engine efficiency by the following:

- The ignition timing can be set closer to the knock limit of the engine, due to more repeatable combustion cycles.

- The band of occurring IMEPs is narrower, thus cycles with lower performance occur less often.

- The lean operation limit of the engine can be increased for a fixed COV_{IMEP} limit, as more margin is available.

- The maximum operating pressure of the engine increases, by a reduction of the $+2\sigma_{p_{Cyl}}$. Therefore the engine can be operated closer at its mechanical limit.

- The exhaust temperature variation is reduced, new OPs are accessible due to a lower mean temperature.

---

[3]per definition of the standard normal distribution

2.1.5 Combustion anomalies

In a SI engine, the combustion is started at a desired moment by a plasma channel between the electrodes of the spark plug. The result is a flame that propagates from the ignition point towards the walls of the combustion chamber. This controlled combustion results in cylinder pressures that are within the range of the cyclic variations of the engine OP. However, circumstances in the combustion chamber can provoke combustion anomalies that can cause engine damage (knocking, preignition) or decrease the performance (misfires).

Knocking

Knocking is a phenomena that occurs after the start of combustion by the spark plug. The unburned mixture in the combustion chamber self ignites hereby locally and propagates fast in the gas nearby explosion limit, additionally accelerated by a pressure wave, before it is reached by the flame front. The events start with an increased ignition advance and therefore define the maximal limit for the ignition, the so-called knock limit.
The fast combustion of a knock event with its rapid pressure increase causes an oscillation of the gas in the combustion chamber, creating a characteristic high frequency on the pressure signal. A high-pass filter on the cylinder pressure signal, makes the oscillations visible and reveals knock events. The amplitude to the filtered signal stands for the intensity of the detonation.
Especially for consecutive knocking cycles, the high pressures that occur can lead to local overheating of engine parts such as spark plug electrodes. The fast pressure rise and the gas oscillation cause additional mechanical erosion in the combustion chamber, mainly on the piston or cylinder head.
With a retardation of the ignition timing after a knock event, the combustion can be stabilized. The knock limit of an engine can be increased by reducing the cylinder gas temperature or the combustion duration.

Preignition caused by surface ignition

If the combustion in the engine is not yet initiated by the spark, but starts earlier in the cycle, it is defined as a preignition event. Preignition is mainly caused by overheated engine parts, such as spark plug electrodes or particles on exhaust valves, which cause an auto ignition on their hot surface. Examples for the cause of thermal overload are consecutive knock cycles or a not adapted ignition timing for a change to a richer air-fuel ratios. Smaller surfaces are more likely the cause of the mixture ignition in the engine as they heat up faster. However, tests show that a larger surface needs lower temperature for a surface ignition [10, 32, 72, 73]. Different materials also react differently with the mixture[4].

---

[4]Coward and Guest showed in their work that the surface ignition temperature of platinum is significantly higher than for nickel [28]

The surface ignition is a self sustaining phenomena and causes a fast engine damage, if the fuel supply is not cut off.

Misfires

A misfiring cycle is defined by a very late or not existing combustion, event though the correct ignition timing is applied. There can be multiple causes for misfires in a SI engine. Mainly the mixture conditions at the spark plug are responsible for an inadequate combustion. Examples are a too high flow velocity or too lean or rich local air-fuel ratios between the electrodes of the spark plug, which retard the energy transformation at the begin of the combustion. Another cause may be a dielectric problem, resulting in an electrical discharge at a different position than the electrode gap between center and ground electrode. This might be a surface discharge over the ceramic tip end of the spark plug, a puncture of the ceramic or a flash over at the outside of the spark plug. The resulting delay of a bad combustion begin cannot be compensated during the following energy transformation and causes a lower cylinder pressure at a later crank angle. Misfires impact mainly the drivability and reduce the performance of the engine. However, the unburden fuel leaves the combustion chamber via the open exhaust valves, potentially causing damage to the exhaust components.

2.1.6 Efficiency potentials that are objectives of a jet ignition and their design impact on the pre-chamber

The introduction of a pre-chamber changes the engine from a single to a multi combustion chamber. The separation of the two volumes addresses to recover different efficiency potentials, mentioned in previous sections: a faster combustion of a higher compressed leaner mixture with less cyclic variations.
The following characteristics of the pre-chamber ignition help to reach these objectives:

- Separation of the flow field of MCC and pre-chamber: The flow field in the MCC can be increased without penalizing the early flame kernel growth at the spark plug or the flow between the electrodes can be increased for a given air movement in the MCC.

- Separation of the mixture composition between MCC and pre-chamber: Especially with an active fuel supply into the pre-chamber, a richer mixture in the pre-chamber can be achieved with a lean mixture in the MCC.

- Increase of the flame front surface area due to the multi point ignition by the hot gas jets.

As a result, a functioning pre-chamber system allows the engine to operate at leaner air-fuel ratios, is less dependent on the turbulence field in the MCC, reduces the cyclic variations and increases the knock limit by its faster combustion.

To recover these potentials, the design of the engine needs to evolve[5]. The design of the pre-chamber ignition system drives and is driven by the combustion chamber geometry. Its volume is often correlated with the cylinder volume at TDC [48]. The increase of the compression ratio decreases this volume and hereby also the targeted volume in the pre-chamber for a fixed ratio. This means a higher compression ratio should be accompanied by a decrease of the pre-chamber volume. The shape of the combustion chamber especially the piston is also often adapted to the torches of the jet ignition to match the compression volume [2, 80, 130].

The orifice design that connects the volume of pre-chamber and MCC should assure a good and steady combustion in the pre-chamber by the air movement of the inflowing gas during the compression stroke. The mixture in the pre-chamber should therefore be richer than in the MCC for an overall lean combustion with a λ above 1 [48].

After the jet ejection, due to the pressure increase in the pre-chamber, a rapid combustion in the MCC should be focused. The design of the pre-chamber orifices can help to direct the flame development to areas where knocking occurs. To reduce quenching of the flame, the piston crown geometry should be adapted to the flame propagation that is caused by the hot gas jets and the increase of the compression ratio.

2.1.7 Emissions and future operation fields of spark ignition engines with jet ignition

Combustion engines in motorsport are mostly not limited by emission restrictions. However, the 2020 Deutsche Tourenwagen Masters regulations prescribe a catalytic converter [33]. Although its function during the race is questionable, as the engines are operated under lean conditions. But as the importance of engine emissions will be continuously growing in the future and the pre-chamber ignition offers possibilities to reduce them, they should be mentioned.

The faster combustion of the pre-chamber reduces the exhaust temperature of the engine. This heat reduction can be used to compensate the full load enrichment in road vehicles. An engine mapping of $\lambda=1$ even at full load is target of recent engine development [43]. Instead of water injection, the pre-chamber could be an alternative to assure the proper function of the three way catalytic converter for all OPs.

A more progressive way to reduce emissions of an ICE is a lean burn engine. Particularly Japaneses manufacturers were developing these engines at the end of the last century, to reduce the fuel consumption of their cars. The main disadvantage of a lean burn engine is that the three way catalytic converter cannot be used. To reach an emission level that fits actual regulations, the engine would need to run at very lean conditions to reduce nitrogen oxide (NO_x) emissions with a stable combustion, to reduce HC emissions, which tend to increase for leaner mixtures. An active pre-chamber with fuel supply could be an asset to realize these OPs [6].

---

[5]For example by an increase of the compression ratio and thus changes of the combustion chamber

2.2 Spark ignition fundamentals

2.2.1 Spark plug structure

The spark plug is one of the most fragile parts in a combustion chamber. The need to use different materials including a ceramic and its assembly method[6] make the spark plug vulnerable towards the severe conditions in an engine.

The insulator assembly is the core of the spark plug, it includes the center electrode and the resistor. Aluminum oxide is generally used for the insulating ceramic, which guarantees the dielectricity that is needed to provoke the spark breakdown between the electrodes inside the combustion chamber. The resistor is made of silicate powder that is mixed with e.g. copper and carbon. During the assembly process this mixture is heated up to its melting point to grant a good connection between the metals of terminal and center electrode and the resistor.

For the center electrode, a rivet of precious metal like platinum or iridium is welded to the nickel electrode, this increases the lifetime of the spark plug as wear is reduced. Ground electrodes are often made from steel or a precious metal if required. The gas tightness of the spark plug is assured by an internal gasket between ceramic and shell. This steel part is also responsible for most of the heat transfer from the ceramic towards the cylinder head via the shell.

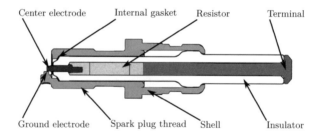

Figure 2.8: Schematic of a spark plug, which includes the shell and the insulator assembly. The latter includes the ceramic insulator, resistor, terminal and center electrode

The internal gasket is deformed by a charge that is applied via a press on the upper part of the spark plug shell while the spark plug shell is locally heated up. Figure 2.8 shows the different components of a regular spark plug.

2.2.2 Spark formation and ignition

The plasma that starts the combustion in a SI engine is formed between two electrodes at voltages of up to $42\,\mathrm{kV}$ for conventional ignition systems [34]. The spark can be divided into three phases: the breakdown, arc and glow phase [1]. Figure 2.9 illustrates the voltage

---

[6]The ceramic is compressed by the metal shell, over-torquing during its mounting or the different thermal expansion of steel and ceramic can cause a loss of the gas tightness

and current characteristics for the phases. The breakdown is responsible for creating the conductive channel, the spark. The formation of the channel is initiated by one electron that is accelerated in the electric field between the electrodes. After a certain free distance of acceleration, the electron hits atoms or molecules in the air. If the kinetic energy of the electron is sufficient, the impact causes an ionization of the collision partner. Thereafter, the new ions are themselves accelerated by the electric field. The number of electrons that are released by this so called avalanche or Townsend discharge can be described by [120]:

$$\gamma \cdot (exp_{\alpha \cdot g} - 1) \geq 1 \tag{2.9}$$

The variables in Equation 2.9 are the third Townsend coefficient (γ), the first Townsend coefficient (α) and the electrode gap (g). Equation 2.10 describes the voltage (U_B) that is required for the spark breakdown [90]. The voltage is hereby dependent on the distance between the electrodes (d), the pressure (p) and γ. The variables A and B are fitting parameters.

$$U_B = \frac{B \cdot p \cdot d}{ln\left(\frac{A \cdot p \cdot d}{ln\left(1+\frac{1}{\gamma}\right)}\right)} \tag{2.10}$$

An increase of air density between the electrodes results in less free distance for the acceleration of the electrons. Thus the breakdown voltage increases. The result of the breakdown is a narrow high conductive channel of less than 50 µm that can port currents ($I_{iA.0}$) of over 100 A [1]. The breakdown does only take about 10^{-8}s, during this time the capacity of the spark plug (C_{SP2} in Figure 2.9) is discharged. The breakdown phase ends when the voltage drops to about 1/10 of its maximum voltage (U_{iA}). The current ($I_{iA.1}$) that is reached during this phase is dependent on the energy capacity of the spark plug (C_{SP1}) and the spark plasma. The arc phase follows the breakdown phase when a hot cathode spot is developed and the discharge is turned into an arc. During this time the capacities of the ignition cable (C_{Cable}) and the high voltage source (C_{Coil}) discharge. Due to the resistance of the spark plug (R_{SP}) and the coil (R_{Coil}), the current ($I_{iA.2}$) is limited to a few ampere. The voltage (U_A) reaches – in dependence to the pressure – between 40 V and 200 V [1]. Because of the lack of time, a stationary arc discharge does not occur in the engine, as this would take several milliseconds. After the capacity discharge in the arc phase, the current is dependent on the internal resistance of the high voltage source. The last phase of the spark is the glow phase with an increase in spark voltage (U_G) and a reduction in current (I_G). During this time the provided electrical energy by the ignition coil (L_{Sec}) is transformed completely into thermal energy [1]. The primary coil with its resistance (R_{Prim}) is then recharged (L_{Prim}) for the next spark.
The spark current and voltage is displayed in Figure 2.9, together with the electrical schematic of the ignition system [98].

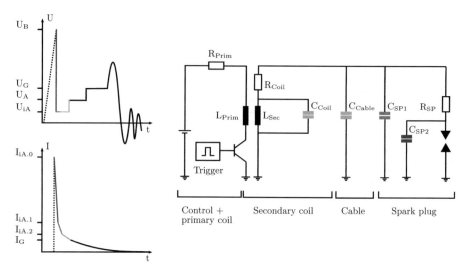

Figure 2.9: Voltage (U) and current (I) of the spark phases on the left side in the Figure. The schematic on the right illustrates the ignition system circuit according to Rager [98] with the different capacities (C), resistances (R) and inductances (L)

Common ignition coils deliver energies up to 90 mJ. The released energy (E_{spark}) during the spark is the integral of the product of the spark voltage (U_{spark}) and the spark current (I_{spark}).

$$E_{spark} = \int U_{spark} \cdot I_{spark} dt \qquad (2.11)$$

The energy that is transmitted effectively to the gas mixture is only a fraction of the supplied spark energy. One cause therefore is the limited time for ignition, especially in engines that operate at high engine speeds, as the duration of the phases are set: breakdown (some nanoseconds), arc (some 100 microseconds), glow phase (some milliseconds) [98]. However, the time for the discharge does not adapt with smaller ratios of s/°CA [7] as it is based on the ignition coil energy. Once the combustion has started, the spark continues in the burned mixture and has no further effect on the combustion.

Another cause is the limited efficiency of the energy transport via the spark.

The losses that occur are divided in relatively small radiation and heat losses, which become substantial during arc and glow phase. The comparison of the resulting plasma energy in Table 2.1, together with the reflection about the limited available time for the ignition are revealing the breakdown and arc phase as crucial for a correct ignition.

If the energy is connected to the available time, the breakdown phase reaches the highest power level (∼1 MW) but consists of only a relatively small amount of energy (0.3 mJ - 1 mJ). The glow discharge energy is the highest (30 mJ - 100 mJ) due to its long duration, but at the lowest power level (∼10 W)[55].

---

[7]e.g. for an engine that operates at 12000 rpm, 1 °CA takes only 0.0139 ms

Table 2.1: Energy distribution and losses in the different arc phases according to Maly [1]

	Breakdown,%	Arc,%	Glow,%
Radiation loss	<1	5	<1
Heat loss to electrodes	5	45	70
Total losses	6	50	70
Plasma energy	94	50	30

Regarding to the different power levels of the three phases, it is not surprising that for a given energy amount the ability to ignite the mixture in the engine differs as shown in Figure 2.10. The breakdown offers hereby the highest potential to ignite lean mixtures.

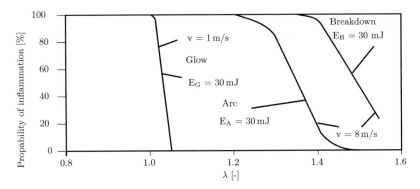

Figure 2.10: Probability of CH_4 inflammation for the different spark phases and different mixture velocities (v) according to Maly [82]

The energy of the ignition coil extends the glow phase of the spark, thus the phase that has the smallest impact on the igniteability of the mixture. For OPs where richer mixture fields are transported to the formed spark, a longer glow duration is beneficial in combination with an increased flow towards the electrodes [78]. However, in full load the engine depends on a repeatable fast ignition that is initiated by breakdown and arc phase. Earlier experiments with a SmartCoil2 system from Apojee show that on a turbocharged race engine even 3 mJ are sufficient to ignite a mixture of $\lambda=1.15$ in full load.

The velocity of the gas between the electrodes has an additional impact on the ignitability of the mixture. Figure 2.11 shows the impact of the velocity on the minimal necessary ignition energy. Close to $\lambda=1$ the required energy has its minimum. If the mixture is deluded, the needed ignition energy increases rapidly. The spark formation is hereby not greatly affected by the air-fuel ratio, contrary to the inflammation process and the thickness and rate of propagation of the flame [55]. The dilution of the mixture results in a decrease of the chemical energy, flame temperature and flame speed. This increases the time for the heat losses towards the electrodes, resulting in an even greater disadvantage as less energy in the mixture is available [55]. A counter measure to this effect can be the reduction of the electrode surface to increase the lean limit of the engine [93].

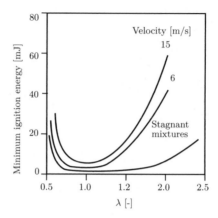

Figure 2.11: Required ignition energy for different mainstream velocities (turbulence = 1 %, pressure = 0.17 bar) according to Ballal and Lefebvre [11]

Different research activities focus on the flow field in the spark gap. The investigated velocities of the air can hereby reach up to 65 m/s [102], although most publications investigate velocities up to 15 m/s [66, 71, 103]. The short breakdown phase is hardly affected by this air motion. The spark deflection by the air flow happens during the arc and glow phase. An larger spark length is beneficial to increase the energy transfer into the mixture and to reduce the heat losses towards the electrodes. However, the energy density decreases as the spark reaches a larger volume.

2.3 Pre-chambers for small volume spark ignition engines

The introduction of a pre-chamber ignition system in an ICE targets an increase in engine efficiency or the ability to operate the engine at OPs that are not inside the fixed criteria of combustion stability or air-fuel ratio. An overview in 2010 about the jet ignition systems for small volumes (<3 % clearance volume) is given by Toulson et al. in [119] where the jet ignition is compared to diesel, SI and CAI engines.

The creation of multiple ignition spots by the ejected jets in the MCC reduces the dependency of the conditions at the spark plug position, as the combustion in the MCC is initiated simultaneously at several spots. As the internal flow inside the pre-chamber can be guided via the orifices, the flow conditions inside the pre-chamber can be better adjusted, compared to a standard spark plug [129]. The mixture in the pre-chamber volume can be separately controlled in an active pre-chamber system with fuel enrichment or a dedicated spray targeting for passive pre-chambers.

After the ignition of the mixture in the pre-chamber by the spark, the combustion creates a pressure difference between the pre-chamber and the MCC. As a result, gas is flowing through the pre-chamber orifices forming the hot gas jets that ignite the principal mixture in the MCC.

The position of the electrode gap in the pre-chamber can hereby influence how much unburned gas is pushed out of the pre-chamber before the flame front reaches the orifices. A lower electrode gap position is also beneficial in regards of burned residuals in the pre-chamber [105].

The diameter of the holes in a pre-chamber are responsible for the formation of the jet that is ejected into the MCC. The discussion about whether the ejected media are flames or extinguished jets can be broken down to the diameter of the hole. Smaller holes quench the flame and products of the unfinished combustion – mainly OH and CH radicals – are then re-igniting the mixture in the MCC. Flame torches are established when the hole diameter is increased, while the number of radicals decreases. Smaller holes increase the penetration of the jets into the MCC, however chocking of the jet needs to be avoided as discovered by Bunce et al. [22].

The term hot gas jets is correct in both cases and is therefore used in this work. The effect on combustion by the jets is mainly an increase of the flame front surface, therefore the number of pre-chamber holes can have a beneficial effect [62]. While a standard spark plug initiates a combustion with a circular flame propagation that starts at the spark plug, a pre-chamber creates multiple starting points in the MCC via its jets. A result is a decrease in burn duration and an increase in combustion stability. Especially in lean OPs, a pre-chamber shows their benefits. The development work of pre-chambers in SI engines focuses often on larger gas engines with a cylinder displacement range of 4 l to 32 l [53, 81]. The adaptation of a pre-chamber system for displacement volumes of less than 0.5 l requires some basic changes. Most obvious are issues due to the limited space in the MCC, as spark plug sizes in road car engines have decreased from M14 to M10 over last decades, while gas engines are using spark plug with thread sizes up to M18.

The pre-chamber can be designed as a separate volume in the cylinder head, which is equipped with a spark plug. For active pre-chambers with a dedicated injector this is generally necessary. Another possibility for a passive pre-chamber is the integration of the volume into the spark plug. The pre-chamber plug can then be screwed directly into the cylinder head. Especially during the development phase, where different designs are tested, this offers a larger flexibility. The use of a direct fuel injection reduces also the available space in the cylinder head.

Most gas engines are used in stationary engine OPs, meaning no requirements for transient behavior and few starts. However, in road cars engine speed changes are permanent and the cold start and other emission relevant OPs are critical. Nevertheless, Honda introduced in 1975 an engine that was able to match emission standards without the use of a catalytic converter, by the use of their Compound Vortex Controlled Combustion pre-chamber system [55].

After a long period without pre-chambers in road cars, Maserati announced in 2020 that their MC20 will have a central mounted pre-chamber and an additional spark plug in the MCC [26]. The pre-chamber is designed as an insert that is directly water cooled. Maserati claims to use the pre-chamber not for a lean burn combustion, but to increase the knock resistance at high boost pressures. At low loads, the engine is operated via the spark plug.

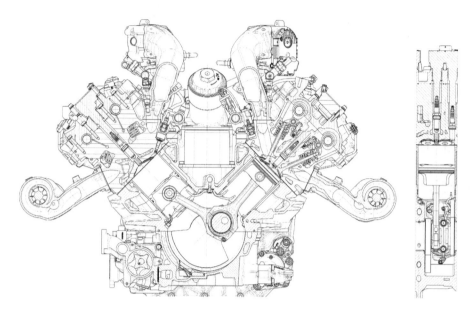

Figure 2.12: Maserati MC20 engine with passive pre-chamber ignition and additional side
mounted standard spark plug [26]

In the patent description the insert is defined by a thermal conductivity of more than
150 W/(m*K) preferably higher than 250 W/(m*K). The wall thickness is announced to
be less than 3 mm for the pre-chamber. The diameter of the holes is defined between
0.8 mm and 1.8 mm, the number of holes is fixed between 6 and 9. The volume of the
pre-chamber should be less than 0,5 % better 0,3 % of the displacement of one cylinder
(which equals 2500 mm^3 or 1500 mm^3 for the 3l Maserati V6 engine) [14].
The Maserati engineers decided to use a passive pre-chamber system, despite the fact that
the engine has already two injectors per cylinder. Both spark plugs are designed differently. In Figure 2.12, the spark plug in the MCC has a standard ground electrode. The
spark plug that is mounted into the pre-chamber has a ground electrode that resembles
to one for race applications. An environment where the spark plug is exposed to higher
thermal load and needs to fit into a combustion chamber with less space in front of the
spark plug, due to the piston proximity by the increased compression ratio. The cold
spark plug and the direct contact to the cooling circuit in Figure 2.12 are showing one
of the main mechanical challenges regarding the turbulent jet ignition: the avoidance of
overheating of the pre-chamber components in full load, to prevent damage of the engine
due to preignition or knocking.

2.3.1 Active pre-chambers

Active pre-chambers use a dedicated enrichment with fuel. Hereby, a small quantity
of fuel is injected directly into the pre-chamber's volume. This might be done by an

additional injector as in the Honda or Mahle turbulent jet ignition (TJI) system [5, 36], a check valve for gas [113] or even an auxiliary intake valve [31]. The objective of the separately controlled fuel supply is to control the air-fuel ratio inside the pre-chamber directly without impacting the mixture in the MCC. The air-fuel ratio in the pre-chamber can therefore be richer than the mixture in the MCC. The cost of additional components is a negative aspect of the systems. Also the additional space that is needed in the cylinder head is challenging. Honda published a patent in 2019 with an active pre-chamber for a motorcycle, showing a possible integration for a small size engine (picture (b) in Figure 2.13).

(a)   (b)

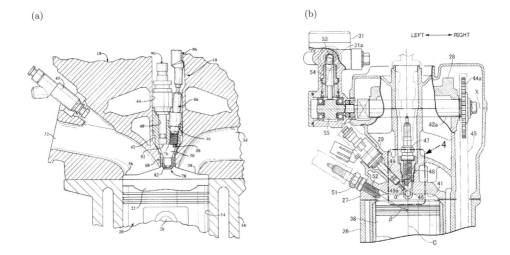

Figure 2.13: Patent pictures of active pre-chamber ignition systems. The Mahle TJI (a) and a Honda system for motorcycles (b) [5, 36]

Publications from Benajes et al. [15] and Sens et al. [105] show that for engines, which are operated with ultra lean mixtures (λ>1.6), an active pre-chamber system is required. The Mahle TJI was developed by William Attard and his team in 2010 and is shown in graph (a) of Figure 2.13 as an example for a concentrated research work on an active system for a road car application [4, 5, 8, 21, 22]. Their development shows the potential of a fuel enrichment in the pre-chamber to increase the lean limit to a λ of 2.2.

The IAV GmbH increased also their research in the last years on pre-chambers and confirmed this limit [105]. In 2018 they introduced an active pre-chamber system, with focus on the reliability challenges in combination with other fuel saving strategies – such as the Miller cycle – to increase engine efficiency [106].

The European H2020 EAGLE project targeted a high thermal efficiency of 50 %. Close to this objective, Serrano et al. published in 2020 a paper about an engine with 47 % thermal efficiency by the use of an active pre-chamber at λ=2.1 [107].

2.3.2 Passive pre-chambers

A passive pre-chamber is dependent on the mixture that enters its volume via the orifices towards the MCC. The design of the engine is therefore less complex, as no additional fuel supply is needed. However, it requires an additional effort for the direct fuel injection development, as the mixture in two volumes needs to be optimized. A positive point of the passive system is the cost aspect, as the pre-chamber can be integrated in the spark plug.

From a development point of view, a spark plug with integrated pre-chamber allows to test different designs more quickly, without disassembling the cylinder head or opening the water circuit as for the Maserati design in Figure 2.12. Also reliability aspects such as wear or deformation of the cap can be addressed more easily by a replacement of the pre-chamber spark plug.

Ferrari published a patent of their pre-chamber concept in 2019 [27], consisting of a pre-chamber insert and an additional spark plug, similar to the Maserati system. However, the position of the spark plug is more central as shown in Figure 2.14.

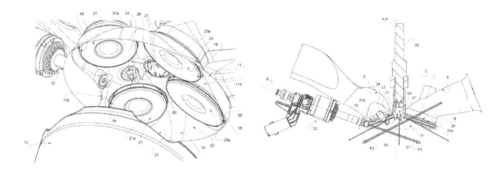

Figure 2.14: Patent pictures of a Ferrari pre-chamber ignition system [27]

3 Experimental setup

3.1 Test bench setup

The different experiments that are the base for this work were performed in France at RSR in Viry-Châtillon and at the IFKM at the KIT in Germany. The test benches are both designed for single cylinder use. Because of the different test benches and engines that are used during the pre-chamber development, it was not possible to use the same fuel for all experiments. Additionally, the engine development at RSR results in various changes of the engine design and architecture over the last years. Permanent improvement of the different components do have a significant impact on OPs and also the ignition system. Therefore, no absolute values are usually stated in the present work, but the differences between the tested ignition systems.

3.1.1 Institute of Internal Combustion Engines at the Karlsruhe Institute of Technology

At the IFKM, the test bench is a container test cell with preparations for multiple measurement devices. The crank angle resolved pressure signals are recorded via a DEWETRON DEWE800. High-pressure signals are measured with Kistler 6054AR piezo-electric quartz sensors. Low-pressure signals in intake and exhaust duct by piezoresistive sensors (Kistler 4045 and 4075). A compressor in front of the engine allows to adjust the inlet pressure. Thereafter, a water heat exchanger conditions the air to the required temperature of 40 °C during the experiments. The exhaust pressure can be regulated via a flap. The air-fuel ratio is measured with two ETAS LA4 Lambda meters and Bosch LSU 4.9 sensors. A Pierburg PLU 401 measures the fuel flow that enters the engine. A Bosch HDEV5 fuel injector is mounted in the intake line to operate the engine with an indirect fuel injection, which results into a homogeneous mixture preparation in the combustion chamber and therefore also in the pre-chamber. Ignition and injection signals are recorded in crank angle resolution. Spark voltage is measured with a Tektronix P6015 high voltage probe, spark current with a Pearson Electronics 2877 current monitor. The test bench is prepared for different optical analyzes such as a camera endoscopy of the combustion chamber. Optical measurement in the combustion chamber is realized with a LaVision HSS6 high speed camera. Engine speed for the experiments at the KIT is set to 2000 rpm. During the experiments at a given OP, the air-fuel ratio is controlled via the boost pressure, while the fuel mass flow is held constant. Engine oil and water temperature are set to 90 °C.

3.1.2 Renault Sport Racing

The single cylinder test bench at RSR was adapted for the needs of the 2014 FIA F1 regulations. Especially the change to a turbo charged system and the focus on engine efficiency made modifications of the test bench necessary. Pressures in the intake, exhaust duct and cylinder are recorded with an AVL Indicom system. The air-fuel ratio is measured with a lambda sensor in the exhaust. Engine speed is set during the experiments to 10500 rpm. Fuel mass flow via the direct injector is determined by the Coriolis principal and set to 16.67 kg/h at full load. Boost pressure is adjusted with an external compressor to the targeted air-fuel ratio. Temperature and humidity of the inlet air can be adjusted individually. Exhaust back pressure can be adjusted by a flap to change the scavenging behavior of the engine.

3.1.3 Virtual test bench

The Development of the pre-chambers is supported by simulation. This includes 3D calculation of temperatures, flow or combustion of the computer aided designed (CAD) components. The finite element simulation requires important computational capacities. To reduce the number of versions that are calculated, a preselection is necessary. For the choice of pre-chamber volumes and the surface area of the orifice holes, 1D simulation has proven to be a useful tool. Even a simplified model with limited input data, such as geometrical engine data and cylinder pressure, can help to narrow down the scope of pre-chamber designs. This reduces the costs for the production of pre-chamber spark plugs and time on the test bench.

The heat release in pre-chamber and MCC are represented by a Wiebe model. Engines at RSR and the IFKM are modeled and the simulated OPs are adapted for the experiments at each test bench. The differences of the OPs[1] make it necessary to build different models for the comparison of the pre-chambers in the experiments. The interpretation of test results is supported by the simulation and the transfer of findings between the test specimen. This counts particularly for the understanding of the gas exchange between pre-chamber and MCC. Test data, such as pressure measurement in the pre-chamber and MCC, are necessary to set up and validate the model.

The spark deflection test chamber (Section 3.4), which is used for investigations of the flow movement in the electrode gap, is also modeled as the flow velocity in the orifices is a relevant simulation output for the analysis of the measured spark deflection.

3.2 Engine data

Experiments at the IFKM are performed on a single cylinder research engine. The cylinder head of the engine is capable to port different endoscopes that allow to capture pictures from inside the combustion chamber from three different angles. The direct fuel injector

---

[1]Principally the engine load and speed and the engines architecture at each facility

and the spark plug are mounted centrally in the cylinder head and parallel to the cam shafts. An additional injector in the inlet port allows engines to run with a homogeneous mixture. Valve timing and lift can be adjusted by a variable valve train. The M12 spark plug thread in the cylinder head allows an easier use of measurement pre-chamber spark plugs. Therefore no modification of the cylinder head is needed for the measurement of pressure and emissions inside the pre-chamber.

The engine configuration that is used for the experiments at RSR is designed to match the technical regulations for the FIA F1 championship. The rules impose the main architecture of the engine such as bore and stroke, number of cylinders and valves, direct fuel injection, maximal compression ratio and different aspects of the turbo charging [40]. The main characteristics of the engines at RSR and the IFKM are shown in Table 3.1.

Table 3.1: Single cylinder engine specifications at the Institute of Internal Combustion Engines and at Renault Sport Racing

		IFKM	RSR
Displacement volume	cm^3	498	266
Stroke	mm	90	80
Bore	mm	84	53
Compression ratio	[-]	10.5:1	12:1
Spark plug thread	[-]	M12x1.25	M10x1.0
Number of valves	[-]	4	4
Fuel injection system	[-]	direct and indirect	direct
Rail pressure	bar	200	200
Maximal mean effective pressure	bar	8	35
Maximal cylinder pressure	bar	100	200

3.3 Pre-chamber ignition systems

During the development of the pre-chamber ignition, multiple pre-chambers are designed and tested both in Karlsruhe and in Viry-Châtillon. All pre-chambers are passive – as prescribed by the rules – and most of them are integrated in spark plugs. The main reason to join the small combustion volume into the spark plug is a faster introduction of the pre-chamber on the race track, as its impact on the cylinder head is reduced. Another reason is the time reduction between iterations, as the pre-chamber assembly can be replaced easily at the test-bench. Different orifice designs or materials can be tested by exchanging only the spark plug.

A pre-chamber insert is another possibility to create the pre-chamber volume. The inset is hereby screwed into the cylinder head and a spark plug is mounted inside. The first pre-chambers were designed as inserts, as the knowledge about spark plug manufacturing was not yet present at RSR.

Investigations in the engine are carried out with M10 (RSR) and M12 (IFKM) spark plugs. The spark deflection test chamber is used for spark plugs with a thread size up to M18.

3.3.1 Passive pre-chamber spark plugs

Passive pre-chamber spark plugs, which were developed during the research, are equipped with different caps that include the orifice design. The installation of the insulator assembly into the spark plug can be done with a nut that assures the necessary pre-charge between ceramic and shell, to guarantee the gas-tightness of the system. By analyzing an originally installed internal gasket and its deformation, the tightening torque of the nut is reverse engineered. The torque is increased until the deformation of an original spark plug gasket is reached. In the case of the present work, the equivalent torque of the nuts (M10 and M12 spark plugs) is between 10 Nm and 15 Nm. The ground electrode is inserted radially into the spark plug shell and forms the electrode gap with the center electrode of the insulator assembly.

An advantage of self manufactured spark plug shells is the increased development speed and a significant cost reduction, as batch sizes can be reduced to one to five pieces. This allows to evaluate different volumes or geometries of the pre-chamber with reasonable cost and manufacturing effort. The pre-chamber volume is closed by a cap towards the combustion chamber. Caps are manufactured as blanks and holes in different directions are then drilled into them, giving the pre-chamber its orifice design. The ground electrode and the cap can either be welded or fixed by a press-in operation into the shell. For a better reliability welding should be preferred.

Principal parameters that are varied during the development are pre-chamber volume, number of holes (n), holes diameter (d), hole offset (o) and the hole angle (β) (Figure 3.1).

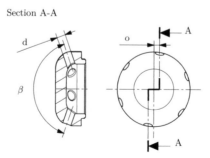

Figure 3.1: Example of the pre-chamber cap parameters: hole diameter (d), hole offset (o) hole angle (β) for a 6 hole design (n)

3.3.2 Pressure measurement inside the pre-chamber

For the pressure measurement inside the pre-chamber during fired and motored cycles an additional sensor must be integrated without impacting the function of the pre-chamber. The sensor that is used is a M5x0.5 cylinder pressure sensor that is integrated into a pre-chamber spark plug with a thread size of M12x1.25. By using the core assembly with a diameter of 4.62 mm from a M8x1 spark plug, the necessary space is found. The core assembly is locked by a nut inside the spark plug shell. For the disassembling of the pressure sensor, the nut must be opened and the core assembly needs to be removed. After this manipulation, the internal gasket between ceramic and shell is changed. Slight differences of the 0.6 mm electrode gap might occur due to an insulator assembly axial position change after re-tightening the nut. For the material of the shell a copper alloy is chosen, to increase the heat transfer from the pressure sensor to the cylinder head. The measured volume in CAD of the pre-chamber is 400 mm^3 including the sensor. The orifice design consists of six holes with a diameter of 0.8 mm and an offset of 0.5 mm. The angle between the holes is 140°. The pre-chamber assembly is shown in Figure 3.2.

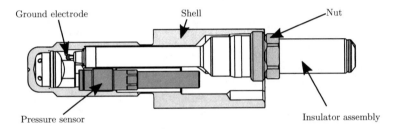

Figure 3.2: M12 measurement pre-chamber spark plug with pressure sensor and removable insulator assembly

For experiments where the M12 measurement spark plug cannot be used, the pressure measurement inside the pre-chamber is realized via a channel that connects the pre-chamber volume to a cylinder pressure sensor inside the cylinder head. To assure that the hole in the cylinder head and in the spark plug are aligned correctly, the channel is drilled with the mounted pre-chamber. It is challenging to reduce the channel length in this configuration to a minimum. However, this is necessary to minimize noise of the pressure measurement and possible failures of the sensor, due to the high amplitude of the pressure signal oscillations [9]. Figure 3.3 shows the pre-chamber pressure measurement in the modified cylinder head.

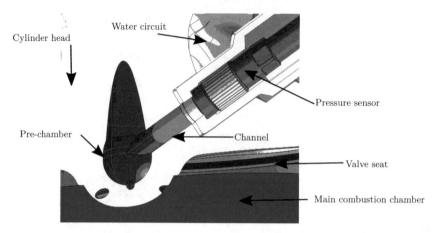

Figure 3.3: Cylinder head modification for pre-chamber pressure measurement with drilled connection channel between sensor and pre-chamber volume

3.3.3 Infrared spectroscopic measurement inside the pre-chamber

A measurement of unburned fuel in the pre-chamber can be realized in most pre-chambers, by replacing the spark plug – or at least the insulator assembly – by a suitable sensor. However, this disables the ignition of the engine and the measurement is only possible for motored OPs. Thus, impacts of the hot combustion chamber walls and their interaction with the spray could not be investigated. The measurement device that was developed during the work at the IFKM, is a fully functional pre-chamber spark plug that allows to compare the fuel measurement inside the pre-chamber for motored and fired OPs (Figure 3.4). Additionally it permits the measurement of burned fuel (CO_2) in the pre-chamber during the fired cycle, and is giving therefore valuable insight about the exhaust gas content inside the pre-chamber. The technique that is used for the determination of the un-/burned fuel is the absorption of IR light of the HC or CO_2 molecules. A method that is used in spark plugs of different publications, to determine local fuel concentration in the combustion chamber [16, 45, 51, 54, 61, 69, 70]. IR spectroscopy is also used to measure local CO_2 in the combustion chamber [44, 52, 88]. However, no publication was found where either burned or unburned fuel was measured inside a pre-chamber.

For the experiments in the present work the ICOS of LaVision is used. An ICOS system consists of two optical fibers, a light source and an IR detector. One glas fiber guides the IR light into the combustion chamber where it is reflected by a mirror and is then captured by the second glas fiber. The space between the mirror and the two cables defines the absorption volume. The second cable guides the light trough band-pass filters to the IR detector. Filter frequencies are $3.3\,\mu m$ - $3.5\,\mu m$ for HC with a wavelength of $3.4\,\mu m$ and $3.6\,\mu m$ - $3.8\,\mu m$ for the CO_2 and water detection.

Relative CO_2 content is calculated by the measured burned gas density and the cylinder pressure, together with an isentropic temperature calculation, based on the measured temperatures at IVC and exhaust valve opening.

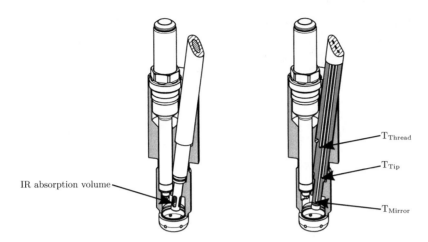

Figure 3.4: M12 Measurement spark plug for IR sensor in the left and temperature measurement positions and device in the right picture. [128]

For the air-fuel ratio inside the pre-chamber, the air mass at IVC and the measured cylinder pressure are calculated with the fuel density from the IR sensor [76]. The sampling frequency of the system is 30 kHz.

The pre-chamber spark plug that is used for the IR sensor (Figure 3.4) resembles the one for the pressure measurement in Figure 3.2. The ICOS is also mounted with its M5x0.5 thread into the spark plug shell. Because of the length of the sensor, the position of the thread inside the spark plug shell has to be inclined with an angle of 7° to bypass the insulator assembly. Therefore, the sensor can be mounted and dismounted without disassembling the pre-chamber spark plug. To confirm that the ICOS sensor is not used in an environment that is too hot, a temperature measurement is performed before its installation. Temperature is measured at three critical points, by using a stainless steel device with type K thermocouples. The revealed temperatures for an 8 bar IMEP OP are stated in Table 3.2. Together with the pressure measurement, operation within the sensor limits can be confirmed.

Table 3.2: Maximal measured IR sensor temperatures and pressures together with the authorized limits by LaVision [77, 128]

		Measured	Authorized
Temperature Thread	°C	156	197
Temperature Tip	°C	197	377
Temperature Mirror	°C	317	-
Maximal IMEP	bar	9	8
Maximal cylinder pressure	bar	50	200

Once the sensor is installed, it is calibrated at λ=1 with the help of the lambda sensor in the exhaust. Between motored and fired OPs the calibration values differ significantly.

The determination of the local air-fuel ratio is based on the air mass in the cylinder and the fuel density (ρ_f). The air mass is calculated with the ideal gas law using the measured p_{Cyl} and the temperature in the cylinder. The cylinder temperature is based on an assumption of an adiabatic compression of the air with the measured manifold temperature (T_{Man}) at IVC. As the limited space in the pre-chamber does not allow an additional sensor, the pressure measurement during the experiment is realized in the MCC.

The difference between the measured pressure and the actual pressure at the sensor position during the compression stroke depends on the orifice design, the engine speed and the pressure level in the MCC. During the pre-chamber experiments pressure differences between pre-chamber and MCC of up to 5 bar were determined[2]. To estimate the possible error due to the lower pressure at the sensor position, the k Factor is introduced:

$$kFactor = \frac{p}{\lambda \cdot \rho_f \cdot T_{Man}} \tag{3.1}$$

The determined k Factor can be used to correct a λ value for a known pressure difference between the measured pressure and the pressure at the sensor position. A setup that occurs when the IR sensor is used in a pre-chamber. The k Factor differs for motored and fired OPs and changes with increasing crank angles.

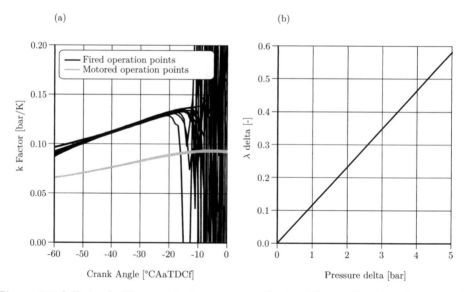

Figure 3.5: k Factor for IR sensor fuel measurement for the different OPs (a) and resulting λ offset for pressure difference between pre-chamber and main combustion chamber (b) with k Factor 0.126 bar/K [128]

Graph (a) in Figure 3.5 shows the evolving k Factor values for the pre-chamber experiment. The lambda delta in the graph (b) shows the air-fuel equivalence ratio difference for a given pressure difference with a k Factor of 0.126 bar/K for the fired OPs. In the

---

[2]An example of the pressure difference between pre-chamber an MCC is shown in Figure 4.39

experiments of the present work the pressure difference between pre-chamber and MCC before ignition was 0.2 bar. The resulting λ correction of 0.03 points was therefore not necessary. However, the k Factor should be applied when a higher pressure difference occurs.

3.4 Spark deflection test chamber

First engine experiments with the pre-chamber spark plug show a significant difference of the spark voltage compared to a standard spark plug. While the spark plug shows the known phases of breakdown, arc and glow, the pre-chamber spark plug has multiple breakdowns and only very short glow phases[3].

For a better understanding of how the pre-chamber should be designed and how the orifices between it and the MCC should be orientated – to optimize the flow between the electrodes – a test chamber is developed (Figure 3.6).

The spark deflection test chamber is connected with a steel tube to the spark plug thread in the cylinder head of the research engine. The connection tube can be adapted to each engine that is focus of the development. It needs to be considered that the additional volume for the test setup reduces the compression ratio of the engine, e.g. 10.5:1 to 8.7:1 for the IFKM engine with a setup volume of $12254\,\mathrm{mm}^3$.

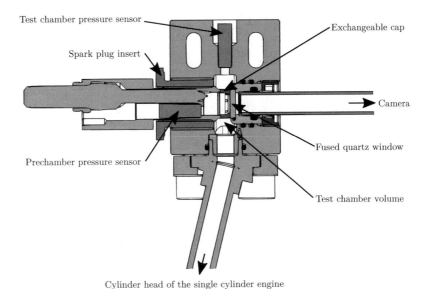

Cylinder head of the single cylinder engine

Figure 3.6: Description of the spark deflection test chamber. The image shows the mounted pressure measurement pre-chamber with an exchangeable cap [129]

---

[3]The difference of the spark voltages can be seen in Figure 4.1

Spark plug are mounted horizontally in an insert that can be adapted for thread sizes up to M18. The end of the spark plug with its pre-chamber is modified, to give axial optical access into the pre-chamber volume, as shown in Figure 3.6. Most of the spark plugs that were used in the experiment are equipped with exchangeable caps. Each cap can hereby port a different orifice design. The open caps are pushed towards a sapphire glass to seal the pre-chamber volume. An endoscope with high speed camera is mounted behind the glass, to record the spark deflection of the motored engine. While the spark deflection measurement in a pre-chamber is a new investigation tool, the optical spark deflection measurement is not unique and was published in other works [89, 91, 103, 108].

During the experiment, the motored engine serves as a pump to reproduce the conditions inside the pre-chamber before ignition. Different pressure sensors inside the pre-chamber, test chamber and MCC give information about the gas exchange between the volumes. The decreased compression ratio, due to the additional volume of the tube and test-chamber, must be compensated by additional boost pressure in order to reach an equivalent maximal pressure in the pre-chamber.

An 8 mm Storz endoscope and a HSS6 high speed camera from LaVision are used for the acquisition of pictures. The crank angle resolution of the camera depends strongly on the pixel x pixel size of the captured image. An area of 320x260 px^2 allows a recording frequency of 48 kHz. This means that for the given engine speed of 2000 rpm every 0.25 °CA a picture can be recorded.

(a)   (b)

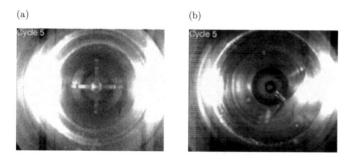

Figure 3.7: Picture of M12 spark plugs in spark deflection test chamber. Multiple spark point plug (a) and single gap spark plug (b) [129]

Figure 3.7 shows the image that is seen via the endoscope of two tested pre-chamber spark plugs in the test chamber. Each picture shows the center electrode in its middle and ends at the pre-chamber wall.

3.4.1 Design of pre-chamber spark plugs in the test chamber experiment

Spark deflection is measured for multiple pre-chamber designs that are listed in Table 3.3. The differences of pre-chambers are the internal diameter and therefore the volume, the electrode distance (ED) for the single gap spark plug (SGSP) and the center electrode design of the multiple spark point (MSP) spark plug. The MSP plug has an extended center electrode that creates four electrode gaps with the pre-chamber wall. Radially mounted pins are the ground electrodes in the M10 and M12 spark plugs. As the center electrode position is fixed by the insulator assembly position in the spark plug, the two different EDs for the M12 SGSP a realized by a radial shift of the ground electrode.

Table 3.3: Pre-chamber spark plugs that are used in the experiment with the spark deflection test chamber

Spark plug	Internal diameter [mm]	Volume [mm³]	Electrode distance [mm]	Orifice design [-]	Breakdown voltage [kV]
M10 SGSP	7	330	0.2	V1	−8.2
M12 SGSP	8	400	0.6	V1-V8	−11.7
M12 SGSP	8	400	1.0	V1-V8	−16.15
M12 SGSP PS	8	400	0.6	V1-V8	NA
M12 MSP	7	330	0.35	V1-V8	−7.4

The spark plug with pressure sensor (M12 SGSP PS) in Figure 3.6 is used for the analysis of the pressure difference between pre-chamber and test chamber and is not optically evaluated. The M12 spark plugs have a common interface that allows to exchange and test different orifice designs. The M10 spark plug has, because of its smaller external diameter, not the possibility to use multiple orifice designs. Its orifice design is therefore directly integrated in the spark plug shell. Tested designs differ in the number of holes (n), hole diameter (d) and hole offset (o), as stated in Figure 3.1. Focus of the different orifice designs is a variation of each parameter independently.

Table 3.4: Orifice designs that are used in the experiment with the spark deflection test chamber

Orifice design	Number of holes [-]	Hole diameter [mm]	Hole offset [mm]	Hole surface [mm²]
V1	6	0.8	0.5	3.02
V2	6	0.6	0.5	1.70
V3	6	1	0.5	4.71
V4	6	0.8	0	3.02
V5	6	0.8	1	3.02
V6	4	0.8	0.5	2.01
V7	4	1	0.5	3.14
V8	4	1.2	0.5	4.52

The designs are shown in Table 3.4. Baseline is the V1 orifice design, V2 and V3 are variations of the hole diameter. V4 and V5 of the hole offset and V6, V7 and V8 have four instead of six holes and different hole diameters.

3.4.2 Spark deflection test chamber methodology

Once the camera has recorded the spark deflection in the test chamber an algorithm post processes each image to transfer the optical measurement into numeric data that can be analyzed. Focus is hereby the spark length and the velocity of its deflection. The sequence from the camera picture to the picture that is used for the calculation are shown in Figure 3.8. The algorithm is programmed in Matlab R2018b. The first step of the picture processing is the spark breakdown detection, the program then uses the algorithm of the following pictures. Picture (a) in Figure 3.8 is an example for a recorded camera picture. The picture is then binarized with an adjustable threshold (picture b). Result is a matrix that consists of ones and zeros for each camera picture. One stands for a white and zero for a black pixel. The threshold that determines whether a gray picture value is white or black, can be adapted to different brightness levels of the camera recording. A higher threshold increases the noise of the images, a lower threshold might not capture the entire spark length. The highest possible threshold is the objective, as it guarantees a maximal spark light detection. Therefore, a filter algorithm is developed to remove the single pixel noise in the image. The filter erases white pixels that are not connected on one of their four sides to another white pixel (picture c). The resulting image can then be processed with the Matlab skel function, where the image is reduced to a 1 px wide line that represents the spark length in pixels[4] (picture d). The calculation from pixel into millimeter is done by a calibration via the measured ED.

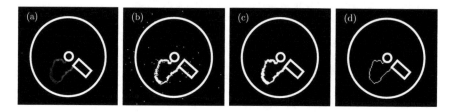

Figure 3.8: Spark deflection calculation methodology. From camera picture (a) over its binarization (b) and filtering (c) to the picture for the spark length calculation (d). Together with electrodes and pre-chamber wall positions in gray color [129]

The calculation of the velocity of the spark deflection (v_{spark}) is based on the length change (Δx) of the spark for a known period of time (Δt). The spark velocity can therefore be determined by the measured spark deflection of two images for a known recording frequency. As the spark follows an U-shape between the electrodes, the maximal deflection can be approximated by halving the spark elongation (Δl_{spark}). Pictures are recorded with

---

[4]The algorithm calculates the length of the spark in pixels and is taking into account if the pixels are diagonally ($\sqrt{2}$ px) or vertically/horizontally (1 px) connected

a constant frequency of 48 kHz resulting in a Δt of $1/48000$ s. This calculation method is also used by Schneider et al. and expressed in the following equation [103]:

$$v_{spark}(t) = \frac{\Delta x}{\Delta t} = \frac{\Delta l_{spark}}{2 \cdot \Delta t} \tag{3.2}$$

The calculation is based on the assumption that the two electrodes do have a fixed position. However, if the spark plug is designed to have a sliding spark contact point (e.g. MSP plug in Figure 4.34), this shift needs to be taken into account. Hereby, the distance of the spark contact point to its origin is added to the spark length for the velocity calculation.

As the camera only records a 2D image of the spark, 3D effects like folding or movement in the direction of the endoscope axis are not evaluated.

4 Results

4.1 Principal difference between engine operation points with and without pre-chamber

A pre-chamber ignition system has a direct impact on the combustion in the engine, as the energy transformation is divided in two stages. The first phase is the combustion inside the pre-chamber, which is initiated by the spark and responsible for the pressure rise in the pre-chamber and the ejection of hot gas into the MCC. The second phase is the combustion in the MCC that starts after its ignition by the hot gas jets.

An ignition of the mixture at different points in the MCC and the hereby increased flame surface decreases the burn duration and makes the system less dependent on flow and mixture inhomogeneities resulting in a better combustion stability.

4.1.1 Impact on the spark formation

Due to the faster combustion, less ignition timing is applied for the same center of combustion[1] for the pre-chamber ignition. As the the plasma formation of the spark plug is later in the compression stroke of the engine, a higher cylinder pressure increases the breakdown voltage and hereby the necessary dielectricity of the ignition system.

However, due to the increased flow velocity between the electrodes and more repeatable conditions at the plasma formation, the distance between the electrodes can be reduced. In the Renault engine, the ED was reduced from 0.4 mm to 0.2 mm after changing to a pre-chamber ignition. The voltage demand during the first breakdown for the same ignition timing is hereby decreased. A look at the spark formation in crank angle resolution shows a significant difference between the standard spark plug and the pre-chamber spark plug. Figure 4.1 shows this impact by a measurement that is performed at the IFKM. The electrode distance of both spark plugs is 0.8 mm. The standard spark plug shows a single breakdown followed by an arc and glow phase, while the pre-chamber spark plug has multiple breakdowns. Both plugs have in common that the first plasma formation takes place during the pure compression pressure in the cylinder. The breakdown voltage is therefore mainly dependent on the electrode distance and the pressure at the spark location. For the pre-chamber the pressure is slightly smaller due to the pressure difference of the orifices towards the MCC. Yet the effect is largely compensated by the later ignition timing for the pre-chamber of 10 °CA to reach the same MFB50.

---

[1]The center of combustion is equivalent to 50 % fuel mass fraction burned point (MFB50)

The spark of the standard spark plug continues after the $-2\,\mathrm{kV}$ breakdown with an arc and glow phase during the following combustion. Inside the pre-chamber the spark forms multiple new plasma channels during the pre-chamber pressure rise after the ignition of the pre-chamber mixture. Therefore, the spark breakdown voltage doubles from $-4\,\mathrm{kV}$ to $-8\,\mathrm{kV}$ (graph (a) in Figure 4.1). This behavior needs to be taken into account during the dielectricity design of the ignition system.

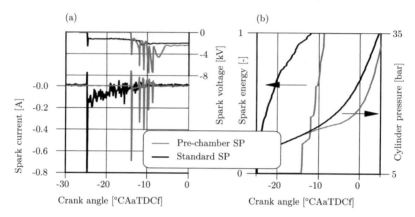

Figure 4.1: Spark voltage and current comparison of a standard and pre-chamber spark plug in (a) and spark energy and cylinder pressure in (b) over crank angle (8 bar IMEP, 2000 rpm). The pre-chamber spark plug has five spark breakdowns while the standard spark plug has only one breakdown. [129]

Spark energy is a product of spark current and spark voltage (Equation 2.11). The time for the energy release is strongly reduced by the use of a pre-chamber as displayed in graph (b), resulting in a possibly faster energy transfer from the ignition coil towards the mixture.

4.1.2 Impact on engine control signals

For a given OP a later combustion initiation with the same MFB50 means a faster pressure rise in the combustion chamber, due to a faster combustion. Hereby, the higher gradient of the cylinder pressure has an effect on the noise level of the cylinder pressure sensor.

In 1986 Spicher and Kollmeier introduced a processing procedure for the pressure signal that helps to characterize the knock intensity in the engine by separating the measured cylinder pressure into a low and high frequency signal [109].

Nowadays, it is common to use a high-pass filter on the cylinder pressure signal, to set the optimal ignition timing in the engine [41]. Kettner discovered that the filtered heat release is more suitable for the knock detection of a pre-chamber combustion [64]. However, this method is not possible for the RSR engine, as the heat release is not calculated in real time on track.

Therefore, the maximal amplitude of the high-pass filtered cylinder pressure signal (Kp) is set as a reference value, which stands for example for a knock event or the desired ignition timing for an OP.

Major impact on the signal has the distance between the sensor membrane and the combustion chamber [9]. However, for the same channel length, the gradient of the cylinder pressure has the biggest impact on the amplitude. The use of a pre-chamber ignition system changes the calibration of the engine by the need to set new higher thresholds. This behavior was also discovered by Sens et al. [105]. Especially in engines that are operated at their knock limit, pre-chamber combustion needs to be controlled differently than with a standard spark plug, as the pressure signal of the pre-chamber could be misinterpreted. Another challenge is the correct analysis of combustion anomalies that are recorded by the cylinder pressure sensor in the MCC. Preignition and knock events in an engine with a single combustion chamber can be well separated by analyzing the cylinder pressure signal. However, the use of a pre-combustion volume adds an uncertainty to the correct interpretation of the pressure signal, as information about the combustion inside the pre-chamber is limited.

The source of inflammation in a preigniting cycle is hereby difficult to divide between a hot spot outside or inside the pre-chamber volume, as the jet ejection is not visible on the cylinder pressure signal of the MCC. A surface ignition that causes a premature combustion in the pre-chamber does not necessarily start the same characteristically combustion in the MCC and the pressure curve in the MCC might only show the symptoms of a knocking cycle.

Figure 4.2 shows the pressure measurement inside the pre-chamber and the MCC of different combustion cycles. Additionally, the 15 kHz high-pass filtered cylinder pressure signal, the knock pressure (Kp) and the current measurement of the ignition coil are displayed. The spark is formed after the current drops to 0 A.

The first two graphs (a) and (b) show cyclic variations inside the pre-chamber for the same OP. The ignition is initiated at $-22\,°\text{Crank}$ angle after top dead center firing ($°CA_{aTDCf}$) in all cycles, but the following pressures inside the pre-chamber are different. The pre-chamber pressure rise of the first cycle (a) is shortly after the ignition and the maximum pressure in the MMC is on a normal level. The second cycle (b) shows an early and fast rise of the pressure inside the pre-chamber right after ignition and a higher maximum pressure in the MCC. Therefore, the noise on the cylinder pressure signal is increased. A filter could wrongly interpret the noise as a knock event [64, 105].

The third cycle (c) represents a standard knocking cycle in the MCC, which could occur likewise with a standard spark plug. The pressure inside the pre-chamber is similar to the non knocking cycle (b). Due to the similar curves of the pre-chamber pressure, it can be assumed that the combustion inside the pre-chamber is the same and the knock event is caused by the conditions in the MCC.

The fourth cycle (d) shows an early preigniton event inside the pre-chamber. The pressure rise at $-45\,°CA_{aTDCf}$ in the pre-chamber indicates the origin of the combustion start. The preignition is initiated by a hot surface inside the pre-chamber, as the pressure rise is earlier than the spark ignition. The ejected gas from $-40\,°CA_{aTDCf}$ to $-50\,°CA_{aTDCf}$ starts the combustion in the MCC, which results in a pressure curve that is almost symmetrical to TDCf.

In the fifth graph (e), the combustion is a mixture between a preigniting and knocking cycle. Due to the comparatively late self-ignition in the pre-chamber, the pressure still has its characteristic shape of a normal combustion but the high oscillation of a knock event in the MCC.

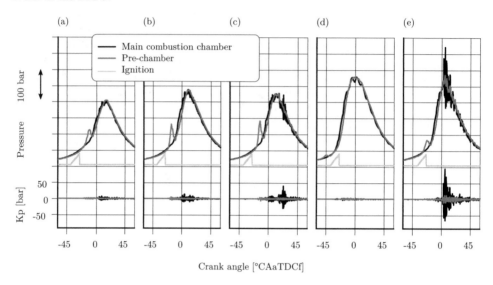

Figure 4.2: Combustion anomalies with pre-chamber and main combustion chamber pressure measurement together with the high-pass filtered pressure signal (Kp) and the current signal of the ignition coil. The graphs show: regular cycles (a) and (b), a knocking cycle in the main combustion chamber (c), preignition caused by surface ignition in the pre-chamber (d) and a preignition cycle together with heavy knock in the main combustion chamber (e)

For the correct control of an engine with a pre-chamber ignition system, especially on its knock limit, the pressure signals must be correctly interpreted and an adequate control mechanism such as ignition retardation and fuel cut-off should be calibrated. If signs for knocking or preignition are not correctly identified, this can cause an overestimation of the maximal temperature capacity of a pre-chamber, as the limit of overheating parts in the pre-chamber might not be detected. The first suspects for the initiation of surface ignitions in a pre-chamber are the electrodes, especially the center electrode. The high temperature due to the limited heat exchange, and the small surface promote an ignition at their surface. Especially for a profound combustion analysis and the mechanical development of a pre-chamber ignition, a measurement of the pressure in both volumes is therefore mandatory.

4.1.3 Impact on torque modulation

The load of a SI engine is generally adjusted by the air that enters the engine, as the range of possible air-fuel ratios is limited by the combustion. The inflowing air is therefore throttled by an adjustable barrier in front of the intake valves, such as barrels, sliders or butterfly valves. The inertia of the air – passing through the system – is significant for turbo charged engines. Especially during gear shifts, the change in engine speed needs an adjustment of the air flow. Additional valves in the intake (pop-off) or exhaust (waste-gate) help to discharge the pressure, to decrease the boost pressure[2]. For the drivability it is therefore beneficial if the engine accepts leaner or richer conditions in the combustion chamber. Examples are the enrichment of the engine mapping before an upshift. The air pressure is therefore reduced before the shift, so that once the new lower engine speed is reached, the air in front of the intake valves is reduced to match the new OP. Another possibility is to lean out an OP with a constant amount of air to reduce the torque or the increase of air for a richer OP, increasing the efficiency and thus the torque of the fuel limited engine. For fast adjustments of the engine torque, the ignition timing is most suitable, as no air or fuel adjustments are necessary. Furthermore, the ignition signal (approx. at $-40\,°\text{CA}_{\text{aTDCf}}$) can be adjusted later in the engine cycle than the injection (approx. at $-380\,°\text{CA}_{\text{aTDCf}}$). The ignition timing range that is available for the torque modulation is between the limits of knock (too early ignition) and misfire (too late ignition). If the ignition system is changed from a standard to a pre-chamber spark plug, this range might change as less ignition timing is applied. It seems likely that the possible ignition retreat is smaller, however the increase in combustion stability of the pre-chamber can compensate this effect. The comparison of the power output and the COV_{IMEP} as a function of the applied ignition timing, with and without a pre-chamber, shows the difference between both ignition systems (Figure 4.3).

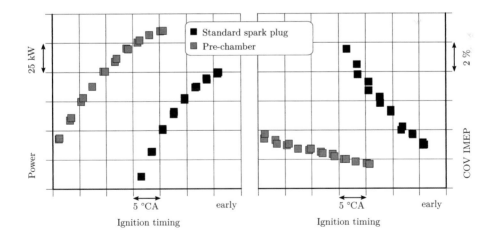

Figure 4.3: Power and combustion stability comparison of pre-chamber and standard spark plug during ignition sweep at 10500 rpm

The tolerated COV_{IMEP} for the engine use is set to 4 %, misfires occur above 7 %. A standard spark plug requires approximately 10 °CA more spark advance than the pre-chamber

---

[2]The use of additional valves in intake and exhaust reduce the overall efficiency, as energy for the compression of air is not used in the engine

to reach the knock limit. The COV_{IMEP} limit of 4 % is attained after the ignition timing has been reduced by 8 °CA, misfires start after a reduction of 14 °CA, which signals the limit of the maximal retreat. The power output can be reduced by 18 % from knock to misfire limit. By the use of a pre-chamber ignition, the power output of the engine is increased. The reason therefore is mainly the faster combustion, additionally cyclic variations are also significantly decreased by up to 8 %. For the same power reduction of 18 %, the ignition timing of the pre-chamber is reduced by 20 °CA, without arriving at the COV_{IMEP} limit of 4 %. But the comparison of both ignition systems at a full load point gives only an indication about potential benefits regarding the drivability, as most torque modeling is required in partial load points. However, the flexibility for the torque adjustment that is gained with the pre-chamber is proven on track as an increase of the usable performance of the power unit.

Drivability of the engine with a standard spark plug depends mainly on the electrode distance. A reasonably bigger distance results in a better drivability, however the probability of failures increases due to the higher voltage demand.

An engine shows a strong dependence on the geometry of its pre-chamber. Different designs change the combustion stability and therefore the possible use of the ignition timing as a tool to change the torque of the engine. Even changes that might be considered of minor impact can change the drivability significantly. An example therefore are the pre-chamber variants in Table 4.1 that all result in a similar power output of the engine at full load. Internal diameter and length of the plug are varied for reliability reasons and thus to increase the lifetime of the ignition plug. The last variant has a central hole in the pre-chamber cap, to investigate the impact of a directed air flow towards the central electrode.

Table 4.1: Pre-chamber plug design variation with different volumes, internal diameters and a design with central hole

Design	Internal diameter [mm]	Volume [mm^3]	Number of holes [-]
A	7.5	363	6
B	7.5	336	6
C	7	316	6
D	7	297	6
E	7	297	7

To compare the different designs regarding their impact on the drivability an engine OP is chosen, which represents an engine speed on a track, where a torque adjustment is required. Therefore the experiment is carried out at partial load (less than 50 % max engine load) and an engine speed of 7500 rpm.

As the engine is not optimized for this low engine speed, the acoustics of intake and exhaust line cause higher gas temperatures in the combustion chamber, which are the cause for a premature knock limit. Together with the COV_{IMEP} restriction, the limits for knock and misfires are close together and the usable operation range is limited. The ignition timing for each plug is varied by 10 °CA from knock towards the misfire limit (Figure 4.4). It can be stated that the combustion stability varies significantly for the different plugs at the earliest ignition timing. A bigger pre-chamber volume shows a better combustion stability. If the volume is reduced, more ignition timing needs to be applied and the combustion stability decreases.

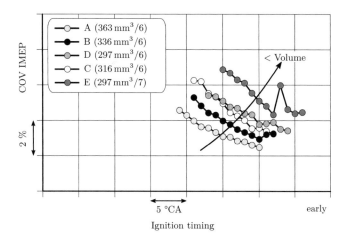

Figure 4.4: Combustion stability of different pre-chamber designs (volume and number of hole variation) at partial load during ignition sweep (sH = 20 g/kg, 7500 rpm)

The additional hole in the center of the pre-chamber cap has a negative effect on the combustion stability. A direct comparison to the six hole version shows higher cyclic variation where even at the maximal ignition timing, the maximal COV$_{\text{IMEP}}$ criteria is not reached.

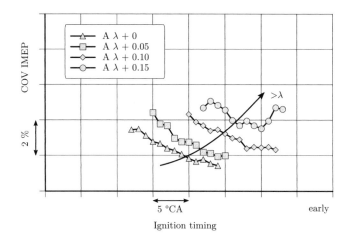

Figure 4.5: Combustion stability of different air-fuel ratios at partial load (Pre-chamber design A, sH = 20 g/kg, 7500 rpm)

The quantity of mixture in the pre-chamber for a given OP is proportional to its volume. However, the quality of the air-fuel mixture – regarding the combustion – is more dependent on its composition. To identify the impact of different mixtures in partial load, the air-fuel ratio is varied for a fixed pre-chamber design (Figure 4.5). The combustion behavior follows hereby a similar trend as during the pre-chamber volume variation. For leaner OPs more ignition timing needs to be applied due to the slower combustion. Combustion stability also decreases as the cyclic variations increase with leaner mixture.

Figure 4.6 shows the comparison of two pre-chamber volumes at different air-fuel ratios and engine speeds. The fuel mass flow ($\dot{m}_{fuel}$) for the OPs at 7500 rpm and 8000 rpm is 31.5 kg/h, which corresponds to about 46 % engine load. The 8500 rpm OP is at 83 % engine load with a fuel mass flow of 68.9 kg/h. The air-fuel ratio is constant for the three OPs. Figure 4.6 shows the increased impact of the pre-chamber design, if the load of the engine is reduced. While at 8500 rpm the combustion stability of the smaller volume variant approaches the level of the better working pre-chamber, a decrease in load reveals the problems that can occur with a design change. This explains also partly why the performance of both plugs at full load is equal, as the impact of a volume reduction is mainly visible on lower partial load. The engine speed itself has hardly an impact, as the comparison of the curves for the 7500 rpm and 8000 rpm shows. An enrichment of the mixture is beneficial as it reduces the combustion stability difference between the two pre-chamber designs (left graph in Figure 4.6).

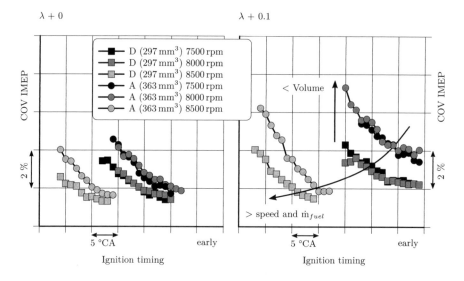

Figure 4.6: Combustion stability of different air-fuel ratios (λ), volumes (V) at partial load with corresponding engine speed and fuel mass flow ($\dot{m}_{fuel}$) and sH = 20 g/kg

During the development of the pre-chamber spark plug, an interaction between humidity and misfires was discovered. An increase of the amount of water in the air resulted in more misfires on the track. Especially in regions with high humidity such as Singapore (absolute humidity about 25 g/m³), the pre-chamber spark plug can cause noticeable drivability problems. To detect possible problems, the experiments at the test bench are carried out with an increased specific humidity (sH) of 20 g water per 1 kg air.

To illustrate the impact of an increased water content in the combustion air, Figure 4.7 shows the experiment with the pre-chamber design at 10 g/kg and 20 g/kg water-air ratio. The air temperature is set to 30 °C and the air pressure to 1000 Pa. These conditions represent a relative humidity of 38 % and 75 %. The effect of the humidity on the combustion stability is similar to a volume reduction (Figure 4.6). As for the volume reduction, the impact is higher at the lower partial load points and at leaner conditions.

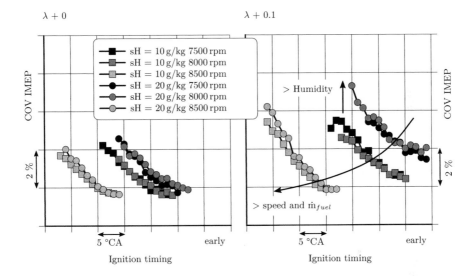

Figure 4.7: Combustion stability of different air-fuel ratios (λ), specific humidity (sH) at partial load with corresponding engine speed and fuel mass flow ($\dot{m}_{fuel}$)

The experiments reveal three major impacts on the combustion stability at partial load when a pre-chamber is used:

- The air-fuel ratio is a partially influenceable value, as in transient engine conditions, the lambda control cannot always provide the target air-fuel ratio. This might be due to the inertia of the air path of the turbocharged engine or the signal lag between measurement, data processing and the resulting injection signal. Misfires do have an amplifying factor, as the unburned mixture causes spikes in the measurement and leads to misguided corrections of the engine control unit.

- Humidity depends on regional conditions and can therefore not be influenced. If the engine needs to operate properly in different regions and seasons, the possible amount of water in the air must to be validated or there should be kept a sufficient margin for a potential combustion stability decrease.

- The pre-chamber volume can compensate an increase in cyclic variations due to humidity or air-fuel ratio. As the only factor that can be directly influenced, the pre-chamber should be designed to allow a sufficient surplus in ignition timing and combustion stability for all relevant engine OPs. The volume is also the only parameter that has no natural limitation such as the saturation of the air with water or the combustion limits of a too lean or too rich mixture.

The impact of each factor is displayed in the main effect plots for the 7500 rpm OP. Therefore, the average COV$_{\text{IMEP}}$ of the first 10 °CA igniton retreat is calculated. The relative strength of the three factors can be compared in Figure 4.8.

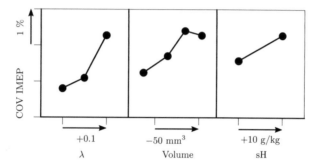

Figure 4.8: Main effect plots partial load (7500 rpm). Base configuration: design D with 297 mm^3, sH = 20 g/kg, λ +0.1

The experiments show the necessity of a pre-chamber validation in partial load with adjustable humidity on the engine test bench before its introduction on track.

Especially the strong impact of the amount of water in the mixture is a non negligible factor that needs to be addressed by an appropriate volume of the pre-chamber or a richer engine mapping. This might also impact the combination of a pre-chamber with a water injection into the engine.

4.1.4 Impact on spark plug thread in the cylinder head

A predictable challenge that occurs when the standard spark plug is replaced by a pre-chamber spark plug in an aluminum cylinder head is the thermal load that is applied to the lower area of the spark plug thread. A standard spark plug temperature decreases from the tip end upwards. However, a pre-chamber spark plug increases the temperature of the lower spark plug thread, as a part of the combustion volume is now inside of the thread. The central area of the combustion chamber in a cylinder head is challenged by the valve seats that are press-fitted into the cylinder head and therefore introduce a mechanical stress towards the spark plug thread. Heating up the area can cause a plastification of the alumina and as a result gripping of the spark plug. Increasing the bore diameter of the thread in the cylinder head and reducing the external diameter of the spark plug helps to assure a proper exchange of spark plugs without seizing. Also the use of an anti-seize paste is recommended.

Additional caution should be taken if a central mounted direct fuel injector is used, as the injector tip – with its gasket – is close to the spark plug thread. The thermal energy that is transmitted by the pre-chamber towards the injector can cause the gasket [3] to overheat. Once the fragilized ring looses the gas tightness towards the combustion chamber, hot combustion gases flow through the opened gap. The result is often the destruction of the injector and the cylinder head, if the leak is not discovered in an early stage.

The water jacket in the cylinder head should be designed for the increased heat that is transmitted via the pre-chamber to its surrounding.

4.2 Test results of different passive pre-chamber designs

The development of pre-chambers for the Renault engine is primarily driven by the need of a mechanicaly stable system that can withstand the thermal conditions in the engine. The material of the shell is hereby key to find a configuration that does not cause preignition during the full load OPs. Once the pre-chamber shell design is fixed, the development of the cap orifice design starts. As the pre-chamber is integrated in the spark plug, different plugs can be compared relatively easily.

4.2.1 Passive pre-chamber with M12 spark plug thread

The first investigations on the IFKM engine are focused on the comparison between a standard spark plug and different pre-chamber spark plugs. 1D simulation is used to determine potential pre-chamber designs. The volume of the pre-chamber, the number and diameter of the holes are hereby varied. To compare the different versions, ejection parameters of the flow through the holes are plotted. Therefore, only the flow conditions during a mass flow from the pre-chamber towards the MCC are considered[4].

Figure 4.9 is showing the Mach number and jet impulse for the four hole design (a and a.1) and for the six hole design (b and b.1) of the hot gas ejection. For a given hole number, a V-shaped area can be determined, for a maximized Mach number of volume to surface ratio for the pre-chamber. For a four hole configuration, the maximum Mach number can be found for a surface-volume ratio of about $0.0023\,\text{mm}^{-1}$. A six hole pre-chamber decreases the ratio to $0.002\,\text{mm}^{-1}$.

If the hole diameter is reduced for bigger volumes, the trapped fuel mass in the pre-chamber is not big enough to create a positive pressure difference towards the main combustion chamber, therefore the Mach number during ejection decreases (dotted area in graph (a) and (a.1) in Figure 4.9). If the hole number is increased for smaller volumes, the trapped fuel energy is not sufficient to create the pressure difference, which is needed for higher Mach numbers in the orifices.

---

[3]The injector gaskets are typically made of polytetrafluoroethylene.
[4]High Mach numbers of the inflowing mixture during the compression in the MCC are not taken into account

Besides the Mach number, the impulse of the jets is of interest. The jet impulse is the integral of the mass flow ($\dot{m}_{jet}$) multiplied with the velocity (v_{jet}) of the jet during its ejection.

$$impulse_{jet} = \int_{\varphi_{ejection_{begin}}}^{\varphi_{ejection_{end}}} \dot{m}_{jet} * v_{jet}(\varphi) \, d\varphi \tag{4.1}$$

The impulse stands for the energy that is provided by the pre-chamber to ignite the mixture in the MCC.

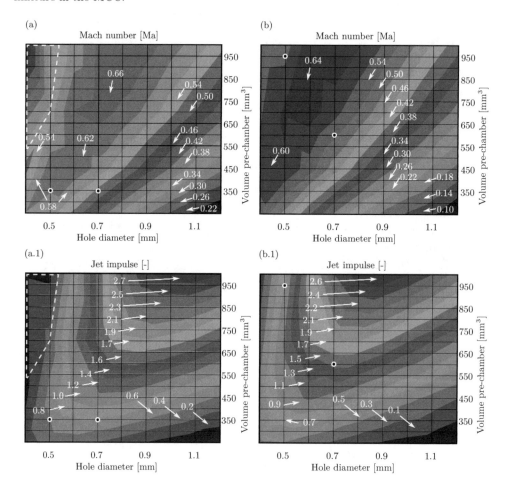

Figure 4.9: 1D simulation results of passive pre-chambers. Mach number and jet impuls for 4 hole pre-chamber (a and a.1) and 6 hole pre-chamber (b and b.1). The white dotted line marks the area with a high negative Mach number of a flow inwards the pre-chamber. The tested pre-chamber designs are marked as white circles in the map

The impulse increases for higher volumes, as more fuel is flowing into the pre-chamber and therefore more energy is available for the hot gas jets. Bigger holes reduce the impulse as the velocity of the jets decreases.

The U shaped areas of the impulse are stretched between the limits of too small holes (little energy in the pre-chamber) and too big holes (little velocity). Jet impulse and Mach number maps for the pre-chamber ejection help to find approximate volume and surface ratios of pre-chambers. For the test at the IFKM additional 3D simulation is used to determine potential candidates for a test in the single cylinder engine.

Table 4.2 shows the four pre-chamber designs that are manufactured, each design is realized with three different EDs of 0.7 mm, 1.0 mm and 1.3 mm. The EDs are predictions from the computational fluid dynamics simulation. It is to mention that the simulation of the plasma and the early flame kernel growth is difficult and therefore the correct simulation of the plasma formation is challenging. Air movement and with it the spark deflection and potential blow outs of the plasma are missing information from the simulation and therefore the predictions are often bigger EDs, as they provide more energy to the early combustion.

Table 4.2: M12 passive pre-chamber spark plug designs

Design	Volume	Number of holes	Hole diameter	Hole off-offset	Orifice sur-surface	Electrode distance
	[mm^3]	[-]	[mm]	[mm]	[mm^2]	[mm]
PC1	600	6	0.7	0.5	2.31	0.7/1.0/1.3
PC2	350	4	0.7	0.5	1.54	0.7/1.0/1.3
PC3	350	4	0.5	0.5	0.79	0.7/1.0/1.3
PC4	950	6	0.5	0.5	1.18	0.7/1.0/1.3
SP						1.0

Different volumes of the pre-chamber saprk plugs are realized by different axial positions of the insulator assembly in the pre-chamber spark plug. The position of the cap inside the MCC stays hereby the same for all plugs. During the experiment in the engine, three of the four pre-chamber designs give useful results. The PC3 design did not ignite the mixture properly and therefore no useful data are obtained. The orifices surface is significantly smaller with this version, therefore the velocity of the inflowing gas is higher compared to the other pre-chambers. Despite the overall poor performance, the smallest ED of 0.7 mm results at least in a few ignitions. The biggest ED of 1.3 mm shows for all designs an unstable combustion for a more deluded mixture. Both observations allow the conclusion that the malfunction of the pre-chambers is based on a blowout of the spark plasma, caused by a too high spark deflection, due to the high velocity between the electrodes.

Pre-chamber spark plugs PC1, PC2 and PC4 are compared to a standard spark plug at 2000 rpm and 8 bar IMEP. The MFB50 is set to 8 °CA$_{\text{aTDCf}}$ for each OP. For the air-fuel ratio sweep, fuel is held constant and air is adjusted via the boost pressure.

Graph (a) in Figure 4.10 shows the comparison of the different pre-chamber spark plugs
at λ=1. After ignition, the flame development angle is approximately 40 % shorter with
a pre-chamber. The combustion duration is also significantly reduced from 40 °CA to
35 °CA. The reduction is hereby caused by a faster early energy transformation from
MFB10 to MFB50 (−42 %). The second half of the combustion from MF50 to MFB90
shows hardly a decrease (−4 %). The performance of the engine is increased by 3 % with
a pre-chamber spark plug. Exhaust temperatures are reduced from 615 °C to 600 °C, due
to the faster pre-chamber combustion. The ED shows no major impact on the combustion
at λ=1. Graph (b) in Figure 4.10 displays the spark voltage together with the pressure
signal in the MCC for the standard spark plug and the PC1 pre-chamber. With its later
ignition timing for the same MFB50 and its impact on the pressure development, the
combustion of the pre-chambers differs compared to the spark plug and shows a later but
faster energy transformation.

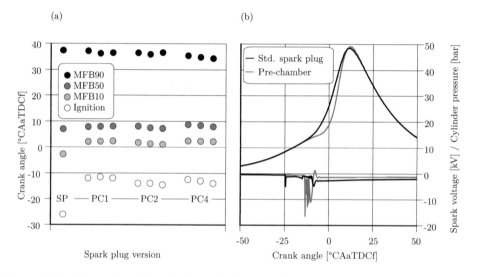

Figure 4.10: Comparison of standard spark plug and different pre-chamber designs. Combustion results are displayed in (a). The cylinder pressure for the PC1 pre-chamber
and the standard spark plug together with the secondary voltage are displayed in
(b)

The combustion stability is comparable for all plugs, due to the rich conditions in the combustion chamber. However, if the amount of air in the chamber is increased, a smaller ED
in the pre-chamber is beneficial as displayed in Figure 4.11. The pre-chamber plugs show
a lower COV$_{\text{IMEP}}$ than the standard spark plug at the OPs where they are functioning
correctly. However, the engine's lean limit is earlier and more abrupt for all pre-chambers,
while the standard spark plug has a progressive degradation in combustion stability with
increasing air-fuel ratio. Smaller EDs help to extend the lean limit of the spark plug, a
controversial observation compared to standard spark plugs [114]. Expectations that the
pre-chamber would allow leaner OPs can not be confirmed during the experiment.

The ranking for the plugs regarding the lean limit is PC1 before PC2 and PC4, which corresponds to the orifice surface of the plugs. An increase of air in the MCC for the diluted OPs means also an increase in velocity of the inflowing mixture into the pre-chamber that causes problems during the spark formation and prevents a proper ignition. The comparison of the spark voltage confirms an increased air movement in the pre-chamber spark plug electrode gap, with multiple restrikes (graph (b) in Figure 4.10). Besides the mixture flow inside the pre-chamber, its composition has an impact on the combustion. The passive pre-chamber depends hereby on the mixture that is formed in the MCC and enters its volume during the compression stroke. In the experiment, the mixture preparation in the MCC is changed by a sweep of the start of injection (SOI) of the direct fuel injector. As a result the air fuel mixture in the pre-chamber changes, too [128].

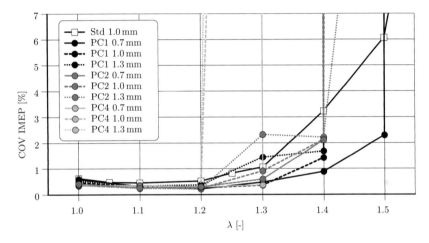

Figure 4.11: Combustion stability during a air-fuel ratio sweep at for different pre-chamber designs and electrode distances compared to a standard spark plug (8 bar IMEP, 2000 rpm)

Figure 4.12 shows the combustion stability for a SOI variation from $-360\,°\text{CA}_{\text{aTDCf}}$ to $-160\,°\text{CA}_{\text{aTDCf}}$, the MFB50 is adjusted by the ignition timing to $10\,°\text{CA}_{\text{aTDCf}}$ and is also displayed in the graph. PC1 and PC2 are showing a stable behavior until $-240\,°\text{CA}_{\text{aTDCf}}$, where the combustion quality drops substantially. The energy transformation with PC3 decreases later at $-200\,°\text{CA}_{\text{aTDCf}}$ but is overall more dependent on the injection timing with OPs close to the COV_{IMEP} limit criteria of $4\,\%$ at SOIs of $-260\,°\text{CA}_{\text{aTDCf}}$ and $-220\,°\text{CA}_{\text{aTDCf}}$.

The standard spark plug is less dependent on the injection timing and allows to run the engine over the total SOI range.

The experiment shows the strong impact of the mixture movement and composition inside the pre-chamber on its performance. Despite the range of the pre-chamber designs, in regards of their volume and orifice design, the performance output differs only little between them. However, they all increase significantly the performance of the engine.

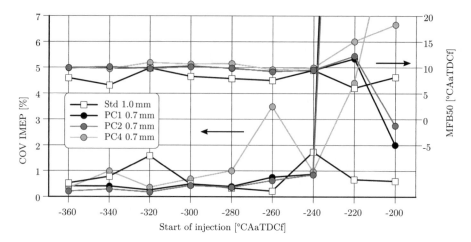

Figure 4.12: Combustion stability during start of injection sweep for different pre-chamber designs compared to a standard spark plug

4.2.2 Passive pre-chamber with M10 spark plug thread

First principal variations of the Renault pre-chamber spark plugs are the hole diameter (d), its angle (β) and the offset (o) of the holes, as shown in Table 4.3. Constant values are the volume of $350\,\text{mm}^3$, the six holes in the cap, the internal design of the pre-chamber and the ED of $0.25\,\text{mm}$.

Table 4.3: M10 passive pre-chamber orifice designs according to Figure 3.1

Design	Hole angle [°]	Hole diameter [mm]	Hole offset [mm]
A	145	1	0
B	140	1	0
C	140	1	0.5
D	140	1	1.5
E	140	0.8	0
F	140	0.8	0.5
G	140	0.8	1.5

The hole diameter is responsible for the pressure drop between the pre-chamber and the MCC. A decrease in diameter results in an increase of the gas velocity and decrease of the mass flow in both directions. The diameter of the six orifices in the experiment is varied from $0.8\,\text{mm}$ to $1\,\text{mm}$. An offset of the holes creates a swirl air movement in the pre-chamber around its longitudinal axis, values are $0\,\text{mm}$, $0.5\,\text{mm}$ and $1.5\,\text{mm}$ for the different spark plugs. The hole angle is connected to the area that is reached by the hot gas jets in the MCC. An increased angle results in less jet interaction with the piston. Figure 3.1 shows a scheme of the different parameters. Results that are compared during the experiment are the brake specific fuel consumption (BSFC) and the cylinder pressure noise level (Kp) of the pre-chambers. The BSFC gives the indication if the pre-chamber performs better from a combustion perspective. A smaller BSFC means a more efficient combustion and a higher power output for a given amount of fuel. Kp stands for the capability to control the engine. The compared noise is the mean amplitude of the high pass filtered cylinder pressure signal. As the ignition timing is set by the help of cylinder pressure sensors, a higher noise level increases the risk to misdetect knocking cycles and also to destroy the pressure sensor due to the higher oscillations in front of the sensor membrane. A lower noise level increases the control-ability of the engine and thus the chance for a fast introduction with the existing engine control unit. Both are primary targets of the engine development under time pressure.

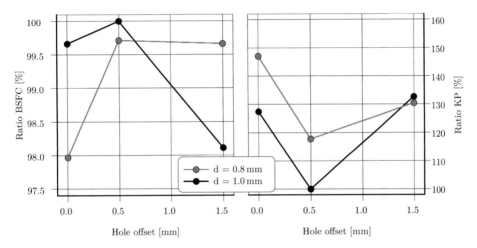

Figure 4.13: Swirl variation (hole offset) of M10 pre-chamber spark plugs with hole diameter (d) of 1.0 mm and 0.8 mm and a hole angle (β) of 140°

In Figure 4.13 the variation of the swirl for both hole diameters is displayed. Spark plugs with the 0.5 mm hole offset show less performance than higher or lower swirl levels. There is a significant difference between the optimal swirl level for the tested hole sizes. 0.8 mm holes work good without a swirl but lose performance with higher hole offsets. 1 mm holes work bad without swirl and increase their performance with a higher swirl. This observation can be explained by the inflowing air velocity that decreases with bigger holes. The offset of the holes provokes a tangential flow inside the pre-chamber and increases the velocity of the air between the electrodes. However, the velocity of the 0.8 mm holes together with the 1.5 mm offset is too high and not beneficial for the early flame kernel growth.

Noise on the cylinder pressure signal is reduced with a smaller hole diameter of 0.8 mm and a hole offset of 0.5 mm. The best compromise for spark plugs with 140° in the tested engine are therefore pre-chamber designs D and E.

As the hot gas is ejected from the pre-chamber close to TDC, there is very limited space between the piston and the cylinder head. Adjusting the opening angle between the holes helps to reduce quenching effects with the piston surface or the cylinder head. However, the angle has also an impact on the noise level. The increase by 5° between an opening angle from 140° to 145° shows a significant performance gain, but also an increase of the noise (Figure 4.14).

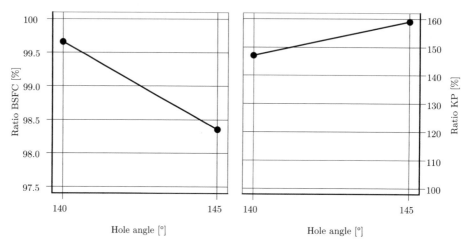

Figure 4.14: Hole angle (β) variation of M10 pre-chamber spark plug with a hole diameter (d) of 1.0 mm and 0 mm hole offset (o)

4.3 Pre-chamber with thermal barrier

Reducing wall heat losses is one way to increase the thermal efficiency of an ICE. Multiple experiments where carried out in the 1980s, to find the theoretical benefit of an insulated combustion chamber [20, 111, 121, 131]. Some of the experiments showed that the increase in wall temperature could not be transferred into engine performance, due to the heating of the fresh gas before the combustion start. However, thin ceramic coatings of about 0.1 mm to 0.5 mm had a beneficial effect in the works of Assanis et al. and Poola et al. [3, 95]. A thermal insulation of the pre-chamber could be of interest due to the smaller trapped air mass, as its temperature has less impact on the MCC combustion and the surface-volume and surface-energy ratios of its small volume are higher than in a MCC. A thermal insulation of the pre-chamber to reduce heat losses is described by Kimura et al., where an increase of thermal efficiency of 0.5 points is measured [68]. Other works show experiments with catalytic surfaces in the pre-chamber and their impact on combustion with an increased wall temperature [59, 101]. Besides a thermodynamic benefit, the insulation of the pre-chamber wall is also of mechanical interest, as the wall is the connection between the cap and the spark plug shell and is therefore exposed to the mechanical stress of the pressure difference between pre-chamber and MCC. A lower wall temperature helps, as the heat transfer towards the cylinder head is reduced and the mechanical properties of the shell material decrease with rising temperature. Especially the limits for hot gas erosion and plastic deformation of the pre-chamber can be increased by an adequate cooling, which is mainly driven by the distance between the spark plug shell and the water circuit in the cylinder head.

From a thermodynamic point of view, there are two controversial mechanisms that are provoked by a hotter cap and pre-chamber wall.

On one hand, a colder spark plug wall and cap exchange less heat with the inflowing mixture. Therefore, the mixture density is higher and more energy is available for the pre-chamber combustion as more fuel is trapped. Another effect of a cool spark plug wall is the reduction of the cap temperature that protrudes into the MCC. If the cap is cooler, the mixture temperature in the cylinder is reduced and the knock limit of the engine can be extended.

On the other hand, a cooler pre-chamber wall and cap increase the thermal losses between the flame and the pre-chamber [68]. Energy that could contribute to the pressure rise and therefore increase the performance of the ignition system is lost to the water circuit.

(a) (b) (c)

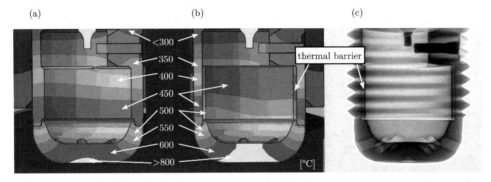

Figure 4.15: Simulation results of pre-chamber with (b) and without (a) thermal barrier and X-ray picture (c) of the manufactured pre-chamber

For the experiment a pre-chamber is equipped with a thermal barrier between the spark plug shell and the pre-chamber combustion volume. The shell of the pre-chamber plug is made of a copper alloy with a thermal conductivity of 300 W/m*K. The wall is covered by a steel tube with a thickness of 0.25 mm and a thermal conductivity of 15 W/m*K. The steel tube heats up with the combustion and is only in loose contact with the pre-chamber walls at ambient temperature. Temperatures of the pre-chamber are calculated via a thermal simulation.

Figure 4.15 shows the X-ray scan of the pre-chamber with the installed barrier and the simulation results. The temperatures are the average value over the engine cycle.

Maximal temperatures – for example the hot temperatures during the combustion – might therefore differ from the simulation. However, the effect of the thermal barrier can be seen in the images. An increase in cap temperature is a result of the reduced thickness of the pre-chamber wall, as the heat transfer is lower due to the smaller cross section.

The insulated pre-chamber is compared in an engine test to the same design without the steel tube. To exclude a possible impact due to a change of the internal geometry, the internal diameter of both spark plugs is the same. Figure 4.16 shows the impact on the combustion for three air-fuel ratios. The flame development angle (Fda) is not reduced with the thermal barrier. No positive effect of the insulation on the early combustion – and with it a possible increase of the ejection energy in the pre-chamber – can be identified. The behavior of the engine is comparable for both spark plugs in therms of knock sensitivity and performance. In very lean conditions the insulated pre-chamber causes a higher knock amplitude (Kp). From a performance perspective, the thermal insulation results in an increase in BSFC of 0.5 g/kWh. It can therefore be considered that the potential benefits of a reduction of the heat losses during the pre-chamber combustion are smaller than the decrease of energy that is trapped in the pre-chamber, due to the heating of the inflowing mixture.

An interesting solution would be the test of thermo-swing material as described by Kawaguchi et al. inside the pre-chamber, to separate the impact of heat loss and density decrease [63]. However, coating of inner walls of a pre-chamber is not trivial and needs further development that is not part of the current development.

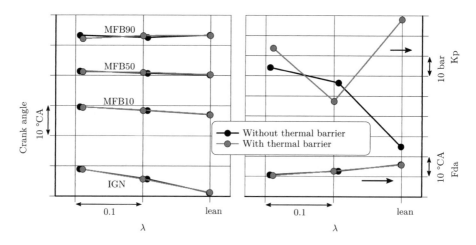

Figure 4.16: Performance of a pre-chamber spark plug with thermal barrier compared to the same design without barrier. The angle of the different mass fraction burned MFB10, MFB50, MFB90 together with the ignition timing in the left graph and the knock amplitude (Kp) together with the flame development angle (Fda) on the right

4.4 Pre-chamber with and without resistor

The energy that is transferred from the ignition coil to the flame kernel is mainly transmitted in the early spark phase (explained in Section 2.2.2). Especially in fast turning engines, the available time to initiate the combustion is very limited[5]. The conditions in

---

[5]For an OP in the Renault engine at 10500 rpm the difference from ignition ($-20.5\,°\mathrm{CA}_{\mathrm{aTDCf}}$) to the peak pressure in the pre-chamber ($-10\,°\mathrm{CA}_{\mathrm{aTDCf}}$) is $10.5\,°\mathrm{CA}$, which corresponds to 0.17 ms. The knock limit is adjusted by ignition timing steps of $0.5\,°\mathrm{CA}$

the pre-chamber determine where the resulting mean effective pressure in the MCC will be situated in the range of possible pressures. Even small variations of the early combustion can result in differences of the IMEP. In addition to optimizing the mixture and the flow conditions inside the pre-chamber, the spark itself has an impact on the initiation of the combustion.

A stable and maximized transfer from electrical to thermal energy during the ignition is desirable. As the ignition itself is initiated during early plasma formation, the focus should be on the maximization of the energy transfer in breakthrough and arc phase.

The spark discharge during the breakdown phase depends on the capacity of the spark plug. During the arc phase, it is the capacity of the ignition coil and the connection between coil and spark plug, which drives the energy. The resistance that is introduced into the system, e.g. in the spark plug, reduces the current during the discharge [23, 79]. Yu measured a significant increase of the spark current from 1.75 A to 20 A when its resistance ins reduced from 4.4 kΩ to 0 kΩ [132]. No resistance would therefore result in more available energy for the combustion begin. However, the resistor helps to reduce electromagnetic noise that could have an impact on the values that are revealed by engine sensors (e.g. cylinder pressure sensors). As the current during the arc phase is also mainly responsible for the wear of the electrodes, its reduction increases the lifetime of spark plugs, as the initial electrode distance is maintained longer [98].

In the presented development both aspects can be neglected, as the Renault engine with its precious metal electrodes does not show a loss of material over the needed lifetime and noise issues are not discovered due to meticulously shielding of all sensors.

The experiment that is performed in the engine compares pre-chamber spark plugs with resistance of 6 kΩ and 0 kΩ. The resistor in the spark plug is hereby replaced by silicate and copper powder, which is melded to assure the gas tightness of the insulator assembly. Both pre-chamber plugs have the same design and are used at the same full load OP, the performance difference can therefore be broken down to the resistor impact.

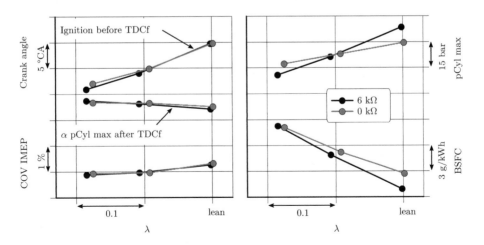

Figure 4.17: Performance of pre-chamber spark plugs with and without resistor during a air-fuel ratio sweep

Figure 4.17 shows the combustion values of both spark plugs at 10500 rpm and different air-fuel ratios. The ignition timing for both plugs is the same, there is no difference in the position of the maximal cylinder pressure (α p_{Cyl} max), therefore no faster combustion initiation occurs. The combustion stability (COV_{IMEP}) is equally for both plugs. In lean conditions the maximal cylinder pressure (p_{Cyl} max) is slightly smaller, causing a performance loss (higher BSFC). The theoretical benefit from a stronger spark on the combustion is not confirmed by the engine test.

4.5 Temperature measurement of the pre-chamber ignition system

The heat transfer from the pre-chamber spark plug to the surrounding cylinder head is important to control the temperature of its components that are in contact with the mixture in the combustion chamber. Overheating can cause engine or spark plug damage, due to loosing of parts like the pre-chamber cap, plastification of the cylinder head thread area or combustion anomalies such as preignition and knocking. Knowledge of the ignition system temperatures helps to prevent these origins for potential damages and is esential for the mechanical design development.

Focus of the measurement is therefore on the temperature of the cap that is in contact with the gas in the combustion chamber, but also on the temperature of the electrodes that tend to overheat. Particularly in the early development phase, the cause of occurring combustion anomalies needs to be narrowed down. The temperature measurement helps to find hot spots in the pre-chamber ignition system and to calibrate the thermal simulation model, which is useful to understand pre-chamber temperatures.

Thermal indicating paint and type K thermocouples are used during the development to determine the temperatures. The MC470-9 from LCR Hallcrest is a paint that changes its color from red over brown to black in 10 color steps. The paint is red until 470 °C and turns black at 1210 °C. Therefore, the painted part needs to be exposed to the hot environment for 10 minutes. For the use on a pre-chamber cap, this means a stabilized full load point during this period of time. Once the test is done, it is recommended to verify the result via a temperature regulated furnace. Therefore, a painted plate is heated up to the estimated color temperature and the colors can then be compared.

The test reveals a cap surface temperature of the pre-chamber in the engine of about 700 °C. The temperature is estimated below the surface ignition temperature of the high octane racing fuel and the cap surface. This is confirmed by pressure measurement inside the pre-chamber and supported by hot surface performance fuel investigations of Davis et al. [32].

However, based on the work of Kutcha and Cato, which revealed a decrease in necessary ignition temperature for larger surfaces [72], and investigations of Coward and Guest about the impact of the material and gas velocity, the cap temperature should be monitored during the pre-chamber development [28].

The limits of the thermal indicating paint are clearly the need of a stabilized OP, the uncertainties about the color interpretation and its application on the surface of a part.

As the paint itself acts as a coating that heats up, the color can only give an approximation about the real part temperature.

To increase the knowledge about the temperature of the pre-chamber system, it is therefore preferred to use thermocouples for a measurement. The acquisition of the temperature over time allows to understand how the pre-chamber reacts on different engine conditions such as air-fuel ratio or ignition timing.

4.5.1 Temperature measurement with M14 pre-chamber insert

Figure 4.18 shows a pre-chamber ignition system including four thermocouples with a pre-chamber copper alloy insert and a standard spark plug. Two sensors measure the temperature of the spark plug, one at the critical area close to the ground electrode and another one at the end of the cylinder head. In the pre-chamber, the cap temperature is measured, as it is one of the parts that tend to overheat. The second measurement point is the wall temperature between spark plug and cylinder head, where the heat is exchanged.

An integration of thermocouples into the pre-chamber or spark plug is challenging, due to the limited space. The use of thin sensors with a diameter of 0.5 mm helps to reach most areas. However, the handling of the plugs – especially the installation – is difficult as the fragile metal tubes break easily during the mounting of the plug. For the installed thermocouples, a contact between the measuring tip and the material should be guaranteed. Otherwise the revealed temperature might be measured too low, as vibrations are causing a movement of the tip between the air and the metal in the cavity. To compare the different cylinders of the V6 engine or different single cylinder engines, this possible measurement error[6] should be taken into account, to decide between real temperature differences or measurement errors. It is therefore recommended to use the same measurement spark plug for different cylinders or engines.

The measurement confirms the hot environment for the spark plug with 600 °C. Based on the measured temperature at the ground electrode, the center electrode can be estimated to be above 1000 °C via simulation and could therefore cause surface ignition. The cap temperature at 760 °C is in a critical range to provoke combustion anomalies. Once the temperatures of the different pre-chamber parts are known, they are valuable information for the choice of suitable materials for the different components.

The contact temperature and the thermal inertia of engine parts define the gradient of its temperature rise when the engine load is increased. The saturation of the parts temperatures might not be reached during the OP. It is therefore necessary to define a protocol for the measurement, to achieve repeatable conditions and to create comparable measurement values. It is also possible to understand transient behavior of the system as function of engine speed and load.

Figure 4.19 shows the measurement of pre-chamber temperatures during two consecutive OPs in a single cylinder engine. The measurement points are displayed in Figure 4.18.

---

[6]It is not unusual that the error of installed thermocouples of different spark plugs is in the area of 50 K.

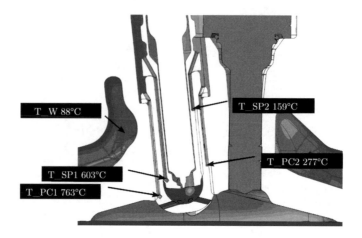

Figure 4.18: Installation of four thermocouples in a pre-chamber (T_PC1 and T_PC2) and a spark plug (T_SP1 and T_SP2). Water temperature (T_W) is set to 90 °C

The temperature close to the MCC (T_PC1 and T_SP1) rises during the 50 s load point with increased ignition advance. The pre-chamber measurement further up in the cylinder head (T_PC2) follows the trend of the exhaust temperature, this can be explained by the close distance to the water circuit, which is heated up by the exhaust gas. The second spark plug temperature (T_SP2) shows only little variation between the full load and idle conditions in between both OPs.

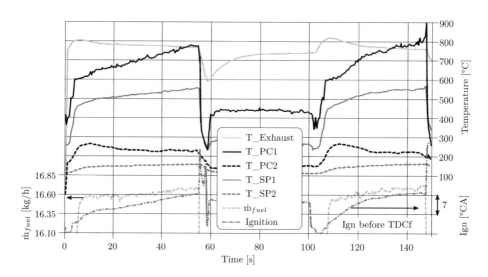

Figure 4.19: Transient measurement of pre-chamber, spark plug and exhaust temperatures together with fuel mass flow and ignition timing.

4.5.2 Temperature measurement with M10 pre-chamber spark plug

The integration of the pre-chamber volume into the spark plug helps to reduce the number of thermal frontiers between the hot insulator assembly and the cylinder head. The lower thermal resistance supports the evacuation of more energy from the spark plug and hence to reduce the critical electrode temperatures.

During the development, this kind of pre-chamber spark plug is also equipped with thermocouples in the cap and at the ground electrode. The experiment with this pre-chamber focuses on OP parameter sensitivities and reveals the impact of the air-fuel ratio in the combustion chamber on the spark plug temperature.

Table 4.4: Engine power, air mass flow ($\dot{m}_{air}$) and pre-chamber temperatures of cap (T_{Cap}) and electrode ($T_{Electrode}$) during a fuel mass flow variation ($\dot{m}_{fuel}$) at different air-fuel equivalence ratios (λ)

T_{Cap} [°C]		λ [-]				
		X-0.1	X-0.05	X	X+0.05	X+0.1
	95	650	(644.5)	639	(633)	627
$\dot{m}_{fuel}$ [kg/h]	100	651	(649)	647	(639)	631
	105	657	(651.5)	646	(641.5)	637

$T_{Electrode}$ [°C]		λ [-]				
		X-0.1	X-0.05	X	X+0.05	X+0.1
	95	256	(255)	254	(252)	250
$\dot{m}_{fuel}$ [kg/h]	100	257	(256.5)	256	(253.5)	251
	105	259	(257.5)	256	(254)	252

Power [%]		λ [-]				
		X-0.1	X-0.05	X	X+0.05	X+0.1
	95	89	(92)	94	(95)	95
$\dot{m}_{fuel}$ [kg/h]	100	95	(98)	100	(101)	101
	105	100	(102)	105	(106)	107

$\dot{m}_{air}$ [%]		λ [-]				
		X-0.1	X-0.05	X	X+0.05	X+0.1
	95	86	(90)	95	(99)	104
$\dot{m}_{fuel}$ [kg/h]	100	91	(95)	100	(104)	108
	105	95	(99)	104	(109)	114

The results are useful to find OPs that cool down the spark plug for a period of time or to avoid OPs that might overheat the plug. Especially when the torque modulation is done with the air-fuel ratio – in addition to the ignition timing – the measurement helps to find parameter combinations at OPs that should be avoided. During the experiment the fuel mass flow is varied by 5 % and the air mass is varied by 10 % around the standard full load OP at 10500 rpm. Table 4.4 shows both plug temperatures as a function of air-fuel ratio and fuel mass flow. Displayed are cap and electrode temperatures, the air mass flow and the power output of the engine. For a better readability the values in 5 % steps of the λ are interpolated and written in brackets. A reduction of the fuel mass flow at a fixed air-fuel ratio has less impact on the pre-chamber temperatures, than an increase of the air-fuel ratio at a constant fuel amount. Especially compared to the power output of the engine, this result is remarkable.

The fuel (vertical $\updownarrow$) is directly linked to the power output of the engine, which means 5 % less fuel leads to 5 % less power at the same air-fuel ratio. Smaller variations might occur due to the different conditions in the engine, which can have a disadvantage for the energy transformation, for example both leaner OPs with 95 kg/h fuel mass flow.

Increased air (horizontal $\rightarrow$) results in a more efficient combustion.

If the fuel mass flow is reduced at a fixed air mass flow (counterdiagonal $\nearrow$), the engine shows less power loss, due to the increase of efficiency of a leaner combustion.

The temperatures of the plug are almost constant in the direction of the main diagonal ($\searrow$) when 5 % fuel and 10 % air is added, even if the power output of the engine shows great differences.

The conclusions about the critical cap temperature of the spark plug are the following:

- $\rightarrow$ An increase of 5 % air and 0 % fuel decreases the cap temperature $(-5\,\mathrm{K})$

- $\nearrow$ An increase of 0 % air and 5 % fuel increases the cap temperature $(6\,\mathrm{K})$

- $\downarrow$ An increase of 5 % air and 5 % fuel increases the cap temperature $(4\,\mathrm{K})$

- $\searrow$ An increase of about 10 % air and 5 % fuel is almost neutral for the cap temperature. $(-1\,\mathrm{K})$

- The spark plug temperatures are not directly connected to the power output of the engine

4.6 Variation of different cap materials

The part of the pre-chamber that separates the pre-chamber combustion volume to the MCC and ports the orifice design is named pre-chamber cap during the development. In the early development stage, the pre-chamber was made entirely out of copper alloy. The high conductive material is necessary to evacuate the heat from the spark plug towards the cylinder head via the threads, Figure 4.18 shows this type of assembly. Less conductive materials cause combustion anomalies like surface ignition, due to overheating of the center electrode or knocking as a result of increased mixture temperature in the MCC, caused by the hot pre-chamber surface.

The lower part of the pre-chamber - the cap - is exposed to a higher temperature than the rest of the pre-chamber. Both sides of the cap and the orifices, which are drilled into the part, are in contact with the hot combustion gas. During the jet ejection, the hot gas is causing a high local heat insertion, because of its temperature, but also due to friction of the fast bypassing media. A pre-chamber that is entirely made out of a single material suffers from the high temperatures that exceed the limit of copper alloys, which starts at about 500 °C [124]. The heat causes an oxidation of the copper on the surface of the parts that are in contact with the combustion media. The erosion of the black copper oxide flakes weakens the part and changes its geometry. Especially the relevant hole diameter might change due to this wear.

As the requirements for the cap material differ from the rest of the pre-chamber spark plug, it is useful to separate the cap from the plug. The cap material can then be selected to suit the needs of its function and environment, without the restrains of the rest of the pre-chamber, for example mechanical requirements of a spark plug thread.

The following three main characteristics of a suitable cap material are fixed for the development:

- Mechanical strength at high temperatures to withstand the pressure difference between pre-chamber and MCC

- Resistance against hot gas corrosion as the cap is exposed to a lean combustion and the traversing hot gas jets

- Thermal conductivity to reduce the cap temperature. This helps to increase the mechanical properties of the material and reduces the probability of combustion anomalies that occur due to the heat exchange between the cap and the mixture in the combustion chamber

Three different materials are chosen to compare their performance and life time as potential cap material: nickel chrome alloy (NiCr), copper alloy (Cu) and nickel copper alloy (NiCu). For a more accurate knowledge of the material, the mechanical characteristics are determined close to the operating temperature in the engine. Yield strength is measured at 600 °C and 800 °C and gives an indication about the strength of the material. A higher yield strength allows more mechanical stress before a deformation occurs. Thermal conductivity for each material is determined from ambient temperature up to 900 °C. As the temperature of the cap is affected by the conductivity and vice versa, the data are helpful for a thermal simulation of the pre-chamber.

Figure 4.20 shows the thermal conductivity and the yield strength of the tested materials. Both NiCu and NiCr alloys show a linear increase of their thermal conductivity over the temperature. Nevertheless, their conductivity is with 40 W/m*K for NiCu and 20 W/m*K for NiCr far inferior to the Cu alloy with 275 W/m*K.

The yield strength shows a significant decrease for the NiCu and the Cu alloy, the NiCr alloy looses only a little of its strength at 800 °C.

The hot gas corrosion resistance is determined after the engine test, as the chemical resistance dependence on the fuel chemistry and the air-fuel ratio in the engine.

The characteristics of the three cap materials is displayed in Table 4.5. Little thermal conductivity, but the best mechanical properties shows the NiCr alloy. The Cu alloy has the highest thermal conductivity but a deficit in corrosion resistance and mechanical strength. NiCu is situated with its strength and conductivity in between the two other alloys. The behavior in the engine determines the limits for each material in regards of the parts temperature and mechanical strength.

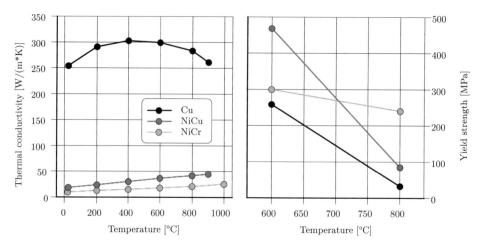

Figure 4.20: Thermal conductivity and yield strength as a function of temperature for the three cap materials

Three pre-chamber spark plugs, each equipped with a different cap material are tested on the single cylinder test bench. The material of the cap is the only difference between the pre-chamber spark plugs, consequently the geometry of the plugs is the same.

Table 4.5: Cap material evaluation matrix

Material	Thermal conductivity	Corrosion resistance	Mechanical strength
NiCr	- -	+ +	+ +
Cu	+ +	- -	-
NiCu	-	+	+

The evaluation of the different materials is based on the ignition timing, the BSFC and the knock tendency of the engine. The ignition timing that is be applied for an OP to reach the knock limit is comparable for the different spark plugs. However, the knock behavior is different for the NiCu and NiCr caps. Both pre-chamber spark plugs respond less on an ignition timing retreat and if the timing is held constant, the knock amplitude (Kp) increases rapidly. The latter can be attributed to an overheating of the cap. The pre-chamber with the Cu alloy cap is easier to control, as it reacts better on the ignition timing.

The comparison is not only about the performance of the plugs, but also about wear over 2 hours in the engine. Therefore, the BSFC of the engine is registered every 1.5 minutes as shown in Figure 4.21.

The pre-chamber spark plug shows no performance advantage for the NiCr and NiCu alloys compared to the Cu variant at the beginning of the endurance test. It can therefore be assumed that the impact of a hotter cap on the ejection energy equals a colder and therefore denser mixture in the pre-chamber.

During the test, the Cu variant loses about 1.5 % of its performance, while both other pre-chamber spark plugs remain on a constant level. The reason for the decrease of the Cu cap can be found during the analysis after the plugs have reached 2 hours in the engine (Figure 4.22).

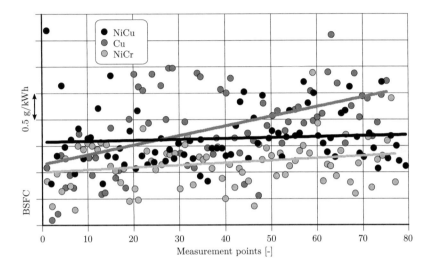

Figure 4.21: Pre-chamber performance evolution during 2 hours full load test. The measurement points are recorded every 1.5 minutes, the trend line of each variant is also displayed

The hole size of the NiCu (a) and the NiCr (b) alloy did not evolve during the test and stayed at 0.8 mm. The optical aspect of both caps is correct without any surface attacks of the combustion gas. However, the Cu alloy (c) shows significant surface oxidation, visible as black flakes. The holes in the copper alloy increased to 0.9 mm (d) due to the corrosion, which explains the loss of performance during the endurance test.

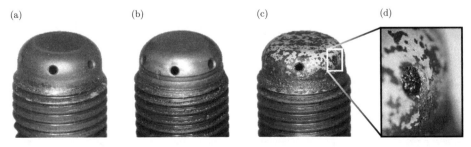

Figure 4.22: Pre-chamber cap surface after 2 hours full load for NiCu (a) NiCr (b) and Cu alloy (c) and (d)

4.7 Variation of orifice designs

The shape of the orifices in a pre-chamber cap are generally cylindrical holes. Researches of Qiao at the Purdue University investigate the impact of different hole shapes on the combustion. The work of Biwas and Qiao shows that a supersonic jet, which is realized by a converging and diverging nozzle, increases the jet performance [17, 18, 97]. The compared values are hereby the jet temperature and the Damköhler Number that is judged as critical for the ignition of a lean mixture in the MCC.

The size of the pre-chamber, including the orifices and the pressure gradient between the combustion volumes differs significantly from the pre-chambers that are focus of the development in the present work (Table 4.6).

Table 4.6: Pre-chamber environment comparison of Renault and Purdue University

		Renault	Purdue
Pre-chamber volume	[mm^3]	<600	>25000
Pre-chamber length	[mm]	<12	>88.9
Pre-chamber diameter	[mm]	<8	>19.05
Orifice diameter	[mm]	<1.2	>1.5
Pressure difference	[bar]	>60	<8
Test setup		engine	test cell
Fuel		gasoline	H_2, CH_4

The orifice designs were tested in a test cell and not in an ICE, with the resulting time limitations for the combustion. The simulation deepness of the Perdue University, regarding the shock formation of the jet after exiting the hole, can neither be realized with the means of RSR nor with them of the IFKM for an environment that corresponds to one of the test engines.

Despite the very different boundary conditions, the results of the Purdue University are the baseline for a spark plug that is tested in an engine. The shape that is chosen to be reproduced and scaled to a M12 pre-chamber spark plug is a converging-diverging (CD) nozzle. The manufacturing of the holes in the pre-chamber cap is generally realized by drilling. The use of a drill imposes a cylindrical hole shape. While the number, position, angle and diameter of the holes can be varied relatively easily, a change of its form is difficult. An additional challenge is the cap shape, as the nozzle design is not drilled into a plate with a plane surface, but into the concave surface of the cap interior. Different manufacturing techniques are investigated to find a solution for the double conical shape. Electric discharge machining and laser drilling seemed suitable, but the cap shape and the hole size discarded both procedures. A 10° cutting tool showed the most promising results and was retained for the production of the pre-chambers. The final nozzle design is therefore partly driven by the possible realization of the nozzle itself.

1D simulation via GT-Power is carried out, to confirm the interest in testing the new design. However, the simulation does not reflect the actual mixture movement inside the pre-chamber and the jet formation towards the MCC. Results in Figure 4.23 are therefore only an estimation about the flow behavior of the nozzle designs in an environment of a combustion chamber.

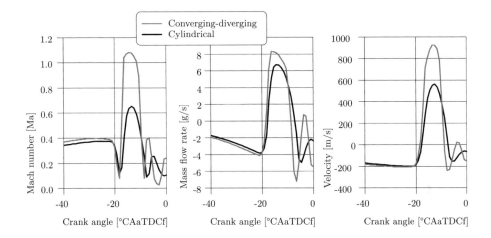

Figure 4.23: 1D simulation results of a CD and a cylindrical orifice design during hot gas ejection at 10500 rpm

A comparison of the CD design with a straight hole shows a significant increase in jet velocity, together with a greater mass flow during the ejection. The Mach number in the smallest diameter of 1 mm reaches also a higher level with the CD nozzle. The positive simulation results encourage the production of the cap design. The integration of the CD nozzle into the M12 pre-chamber can be seen in the X-ray picture in Figure 4.24. Relevant dimensions are measured from the scan. Hereby, the resolution of ±0.03 mm needs to be considered.

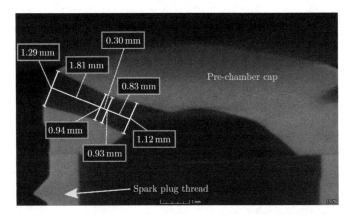

Figure 4.24: X-ray image of the CD nozzle in the pre-chamber cap with measurement of the relevant nozzle dimensions

The engine experiment is carried out with different air-fuel ratios at an engine speed of 10500 rpm. The CD nozzle design is compared to a cylindrical 1 mm hole. Both spark plugs have the same design, except of the orifices in their cap. Therefore, a change of the performance can be connected to the new orifice design.

Despite of the positive simulation results, the test in the engine shows a significant performance loss with the CD nozzle design. The BSFC is increased by 1.65 % at the three different air-fuel ratios. The spark plug requires 4 °CA more ignition advance to reach the knock limit of the engine. The combustion stability is decreased by 0.25 % and the exhaust temperature is 10 K higher compared to the cylindrical hole. All are indications for a slower combustion with the CD nozzle design.

4.8 Spark deflection test chamber results

The results of the previous experiments with passive pre-chambers (e.g in Section 4.2) indicate the impact of the internal flow in the pre-chamber on the ignitability of the mixture. Simulation and experiments of Roethlisberger show the same observation for smaller orifice diameters, causing ignition failure [100].

The measurement of the spark voltage helps to determine, whether the spark is still correctly formed or if its velocity between the electrodes is too high [108, 112]. The connection between the spark voltage and its length can also be drawn [129]. The spark formation in the pre-chamber depends on its internal geometry. As the flow formation inside the pre-chamber is typically a swirl around the spark plug axis, the diameter of the pre-chamber, the electrode distance and its position are important for the correct plasma development. The flow inside the pre-chamber can be adjusted by the orifice design. A decrease of the orifice surface via smaller holes or a reduction of their number increases the velocity of the inflowing gas but has also an impact on the jet ejection via the MCC. A change of the hole offset can increase the swirl level in the pre-chamber without impacting the jet ignition towards the MCC, as the offset is negligible compared to the bore size of the cylinder. The correct design of the pre-chamber orifices should therefore focus on the MCC combustion, but also on the internal flow, as the air movement between the electrodes is crucial in the beginning of the combustion. If air flow is little, more heat of the flame kernel is transferred to the electrodes. Quenching or at least a reduction of the energy that is available for the early combustion phase – in particular for diluted air-fuel mixture – is the result [93, 94]. Most engines operate with velocities of up to 10 m/s, literature states an optimum of 8 m/s [78]. If the air flow is too high, the early flame can be quenched due to the amount of fresh mixture that arrives [102].

The air motion around the spark plug in an engine without a pre-chamber is driven by the tumble or swirl movement in the combustion chamber or by squish effects that occur between the piston and the cylinder head close to TDC. For engines with direct fuel injection, the spray may also create local turbulence and impact the flow field between the electrodes with a late second injection [67]. However, in an engine with a pre-chamber, the air movement at the electrodes is driven by the inflowing gas via its orifices. The hole orientation guides the air flow and creates the turbulence inside the pre-chamber.

Therefore, a separated turbulence level at the early flame kernel development and later for the main combustion in the MCC can be designed. Thus the turbulence in the MCC can then be increased by e.g. high tumble inlet ports, without risking extinguishing the flame at its early state.

A design of the pre-chamber interior is mainly driven by the desired volume and the electrode positions. The center electrode is surrounded by the ceramic insulator and therefore principally in the axial center of a passive pre-chamber. The ground electrode is connected to the spark plug shell. It is possible to use multiple ground electrodes that create air gaps with the center electrode. Common designs for these gaps create them between the cylindrical center electrode and the ground electrodes, which overcome the distance towards the spark plug shell.

The spark plug in an engine is generally exposed to the gas flow from the intake side. If the air gap is not masked by a ground electrode, its position is less combustion relevant [93]. However, in a pre-chamber spark plug with swirl movement, the gas velocity changes in radial direction. While there is only little air movement in the axial center of the plug, it increases towards the spark plug shell. Figure 4.25 shows schematically the vectors of the mixture around the electrode gap.

Spark plug in main combustion chamber   Spark plug in pre-chamber

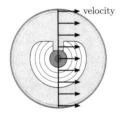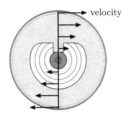

Figure 4.25: Schematic comparison of the flow field at the spark gap in a main combustion chamber and inside a pre-chamber with swirl mixture movement due to a hole offset

The design of the ground electrode can help to guide the air flow towards the electrode gap in the pre-chamber [57].

It is also possible to change the design of the center electrode, so that the electrode gap is formed further towards the spark plug shell where higher gas velocities occur.

To measure the air movement and with it the spark deflection inside the pre-chamber, the spark deflection test chamber that is presented in Section 3.4 is developed.

The pre-chamber spark plugs of the experiment are stated in Table 4.7, a more detailed description of each orifice design can be found in Table 3.4. V1 is the baseline configuration that is also used for the M10 spark plug. Designs V2 - V8 focus each on a parameter change of diameter (d), offset (o) and number of holes (n). All pre-chambers are evaluated optically by the use of a high speed camera, which records an image of the deflected spark every 0.25 °CA at 2000 rpm. The data are then calculated via an algorithm in the spark length and velocity (Section 3.4.2).

A measurement of the pressure in MCC, test chamber and pre-chamber for each orifice design is displayed in Figure 4.26.

Table 4.7: Characteristics of the orifice designs that are tested in the spark deflection test chamber

Design	Characteristic	simulated inflow velocity in the holes [m/s]
V1	Basic configuration	60
V2	Smaller hole diameter	105
V3	Larger hole diameter	40
V4	No swirl	60
V5	Increased swirl	60
V6	Less holes with same diameter as V1	90
V7	Less holes but equivalent surface as V1	60
V8	Less holes with even larger diameter	43

The maximal pressure in test and pre-chamber is higher and later compared to the pressure in the MCC. The reason therefore is the tube which connects the test chamber with the MCC. As the volume of the test setup is directly connected to a reduction of the compression ratio, its minimization is an objective to keep comparable maximal pressure levels during the experiment. The connection tube has a diameter of 6 mm and was optimized with 1D simulation to increase the pressure in the test chamber due to acoustic effects of the gas flow between both volumes. Despite the compression ratio reduction from 10.5:1 to 8.7:1, only 1.1 bar absolute pressure in the intake is needed, for the same maximal pressure level in the pre-chamber as without the additional test chamber volume.

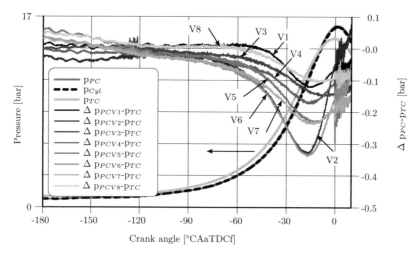

Figure 4.26: Pressure measurement in the different volumes of the spark deflection test chamber together with the pressure difference between test chamber (TC) and prechamber (PC) for the eight different orifice designs [129]

The pressure difference that is created by the different orifice designs shows a maximum of 0.35 bar for the designs with the smallest orifice surface. The absolute level of the pressure is compromised by the sensor tolerance and the fixed point zero-line detection of the piezoelectric measurement. Values in Figure 4.26 are therefore only an orientation for the expected pressure differences during the compression. The noise on the pressure signal at TDC is caused by the ignition.

Besides of the M10 spark plug, three M12 plugs are optically analyzed in the experiment. Two M12 spark plugs differ only in the distance between their electrodes. These single gap spark plugs (SGSP) have an electrode distance of 0.6 mm and 1.0 mm. The third M12 spark plug has multiple spark points (MSP). Four gaps of 0.35 mm are formed between an extended center electrode and the spark plug shell. Both M12 designs are displayed in Figure 3.7.

The spark in the pre-chamber shows multiple changes between the three spark phases of breakdown, arc and glow. Due to the flow field, the spark is stretched during the glow phase. After being deflected to a certain length – during the experiments a maximum of approximately 5 mm – restriking occurs. This chain of events is reproduced until the energy of the ignition coil is used up. The spark voltage in the glow phase with its stable plasma is equivalent to the spark length.

Figure 4.27 shows the spark formation in the pre-chamber for the M10 spark plug (orifice design V1). The graph shows the connection between spark length, spark voltage and the recorded camera picture in the test chamber.

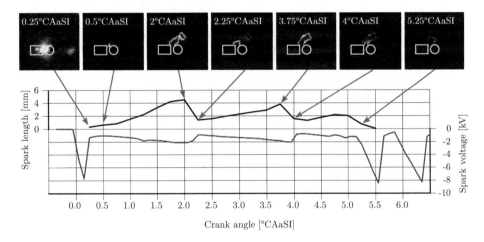

Figure 4.27: Spark formation of the M10 pre-chamber. Spark length together with spark voltage and pictures of the spark during its deflection [129]

The highest voltage of −8 kV occurs at 0.25 °CA after spark ignition (°CAaSI) during the spark breakdown. The spark length of 0.2 mm is hereby the distance between the electrodes, the camera records the high light emissions of the plasma formation. Shortly afterwards in the arc phase, the voltage is reduced to −1 kV. Thereafter the stable plasma of the glow phase is stretched up to 4.5 mm at 2 °CA$_{aSI}$ by the flow and the spark voltage increases to −2 kV. The first restrike event happens at 2.25 °CA$_{aSI}$, where a new plasma channel is formed. The according picture still shows the light of the old plasma path. A second restrike occurs at 3.75 °CA$_{aSI}$ before the end of the spark at 5.25 °CA$_{aSI}$.

4.8.1 M12 Single gap spark plug

The first comparison of the different orifice designs is made with the M12 SGSP and an ED of 0.6 mm. For the experiment the deflection during five cycles is recorded. The spark length of each cycle is displayed in gray in Figure 4.28, the lines of the average cycles are black. The calculated deflection velocity is added as a dotted line. The orifice design specification is indicated in the upper right corner in each graph. For a given hole offset (baseline design V1), an increase of the inflow speed (reduction of the orifice surface, due to a smaller hole diameter or less holes) creates a greater deflection velocity and spark length. This shows the comparison of V2 and V6 with V1. The same effect can be achieved by an increase of the hole offset, as with orifice design V5. Without an offset of the pre-chamber holes, almost no spark deflection occurs (design V4). Low inflow velocities (simulation results of 40 m/s) also show hardly an increase of the spark length (design V3 and V8). For the same simulated velocity (60 m/s) and hole offset a design with six holes doubles the deflection velocity compared to the four hole design (V7).

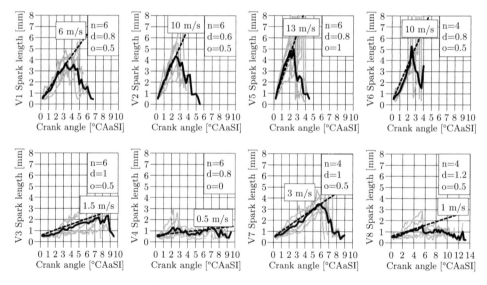

Figure 4.28: Spark deflection for the eight cap designs of the M12 SGSP pre-chamber with an electrode distance of 0.6 mm. The design parameters are stated in the upper right corner of each graph (n = number of holes, d = hole diameter, o = hole offset). The deflection velocity is written in each graph and displayed as a dotted line [129]

The impact of the ED is investigated with two M12 SGSPs. By moving the ground electrode radially to the outside of the pre-chamber, the distance between the electrodes is increased from 0.6 mm to 1 mm, for the same shell design. A comparison of the average spark deflection of five sparks is displayed in Figure 4.29. The 0.6 mm ED design is represented by solid lines, the 1.0 mm distance by dotted lines. The difference is visible at the spark breakthrough when the spark length equals the ED. The upper graph in Figure 4.29 shows the designs with a higher spark deflection (orifice design V1, V2, V5 and V6). When a strong flow field exists in the electrode gap, the results show a larger prolongation for the smaller ED. Especially the orifice design with the bigger hole offset (V5) works significantly better with the 0.6 mm ED, with a 76 % increased spark elongation.

As the smaller ED extends the spark duration, therefore the longest sparks and spark duration is represented by solid lines. This can be explained as the available energy in the ignition coil lasts longer and therefore more time is available for the spark deflection. A larger ED has no positive effect on the gradient of the spark deflection (the spark deflection velocity), even though more air flows through the electrode gap. For little air movement in the electrode gap, a larger ED increases the maximal spark length (designs V3, V4, V7 and V8 in the lower graph of Figure 4.29). The reason is the larger flow field that is caught by the spark, thus a larger air mass can deflect the plasma.

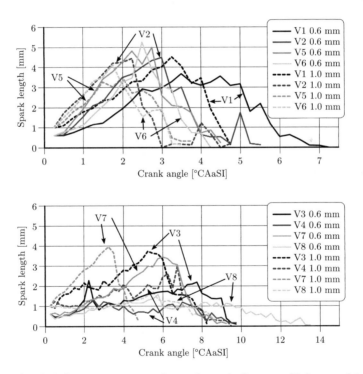

Figure 4.29: Spark deflection comparison for an electrode distance of 0.6 mm and 1.0 mm. The orifice design with a higher deflection (V1, V2, V5, V6) are grouped in the upper graph [129]

As the spark formation depends on the combination of the distance and flow velocity between the electrodes, a more detailed look is helpful to understand the mechanism for a fixed shell design. The comparison of the baseline orifice design (V1) and a reduction in the hole diameter (V2) for both EDs is displayed in Figure 4.30. The curves represent the average of five cycles for each configuration. Besides the spark deflection, the spark voltage is added to the graph. During the breakdown, the larger gap requires a higher voltage as known by Paschen's law [90].

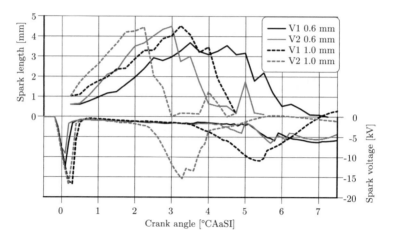

Figure 4.30: Spark deflection and voltage of orifice design V1 and V2 for an electrode distance of 0.6 mm and 1.0 mm [129]

During the arc phase the voltage drops to $-1\,\mathrm{kV}$ for both EDs. This can be explained as the voltage demand in this phase is mainly driven by the pressure, material and the current density [85], parameters that are equal for both spark plugs. During the following glow phase, the voltage is proportional to the spark length.

The high deflection of the V2 design in combination with the 1 mm ED shows a significant increase in voltage from $1\,^{\circ}\mathrm{CA_{aSI}}$ to $2.5\,^{\circ}\mathrm{CA_{aSI}}$. The gradient of the spark deflection is connected to the orifice design and does not depend on the ED but on the the hole diameter. Design V2 with smaller holes shows a higher spark velocity. The maximal achieved spark length is 4.5 mm with higher spark velocity, this value is also achieved by the smaller ED $0.75\,^{\circ}\mathrm{CA}$ later in the cycle. For a lower air velocity (V1 design) the bigger gap results in a higher deflection until 4.5 mm. The smaller ED does not achieve this level but has a $2\,^{\circ}\mathrm{CA}$ longer spark duration.

For the M12 SGSP it can be stated that the gradient of the spark deflection is comparable for the spark plugs with the same orifice design and is therefore independent from the electrode distance. The plasma can be deflected to a maximum of 5 mm, its duration is shorter with larger EDs, as the energy is limited to 55 mJ by the ignition coil.

In the main effect plot in Figure 4.31 the impact of the different orifice parameters on the spark length and velocity are compared. The baseline is the 0.8 mm six hole V1 design with a hole offset of 0.5 mm. The variation in one plot stands for the change of one parameter. The asymptotic spark length of 5 mm can be found in the plot. The graph shows the negative impact of the greater ED, when the offset is increased. A stronger impact of a larger ED for smaller air movements (1.0 mm diameter or 0 mm offset) is also visible.

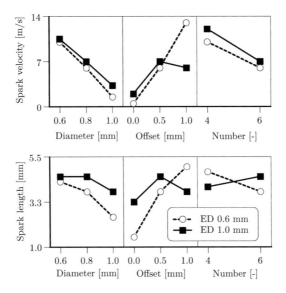

Figure 4.31: Main effect plot for the M12 single gap spark plug for the variation of hole diameter (d), hole offset (o) and number of holes (n) [129]

4.8.2 M10 Single gap spark plug

As space is limited in the combustion chamber, smaller spark plugs are often preferred to
increase the mechanical stability or performance of the cylinder head. This is achieved
by bigger valves or more material between valve seat and spark plug.

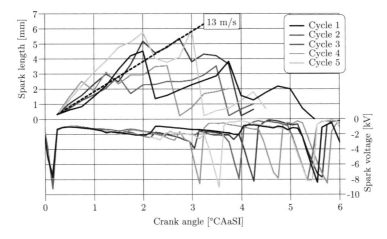

Figure 4.32: Spark deflection and voltage of the M10 V1 design for five cycles, together with
the average deflection velocity [129]

The impact of a smaller spark plug on the air movement inside the pre-chamber is not negligible as its wall diameter is reduced. In the present work the internal dimension decreases from 8 mm (M12x1.25) to 7 mm (M10x1).

The cap of the M10 SGSP ports the V1 orifice design, the shell design is similar to the M12 plug, as the same insulator assembly and ground electrode are used. The position of the ground electrode is set to an ED of 0.2 mm and thus smaller as the 0.6 mm and 1.0 mm of the M12 plugs. Figure 4.32 displays the spark deflection of five cycles. The average deflection velocity is calculated with 13 m/s and is therefore much higher than the 6 m/s of the M12 spark plug. As both EDs in the M12 design show no effect on the gas velocity, the impact of the smaller ED can be neglected (Figure 4.30).

A higher gas velocity in the electrode gap can therefore be directly connected to the smaller internal diameter of the pre-chamber. The maximal spark length for the M10 design is measured with 6 mm, which corresponds to 30 times the initial spark length at breakdown. The average voltage demand during the spark breakdown is -8.2 kV, which fits the 0.2 mm ED.

4.8.3 M12 Multi gap spark plug

The flow field in the symmetrical pre-chamber differs significantly from the axial center towards its wall (Figure 4.25). Simulation shows that the highest velocity can be found close to the spark plug wall [12, 50, 126]. This effect is used for pre-chamber spark plugs, where electrode gaps are formed between an extended center electrode and the spark plug shell [74, 75, 83].

Multiple electrodes augment the lifetime of the spark plugs, as more electrode material can be worn up before the initial ED increases. The design of the MSP spark plug compared to the SGSP is shown in Figure 3.7.

The algorithm (explained in Section 3.4.2) calculates the longest sparks for the V2 and V5 design, which is the same result as for the SGSPs. Figure 4.33 shows the evaluated algorithm results of the orifice design variation with the MSP spark plug.

There is hardly a difference between the SGSPs (Figure 4.29) and the MSP spark plug for all orifice designs, even though the flow velocity in the ED is judged significantly higher for the latter.

The reason for the similar spark deflection length is the changing contact point of the spark on the pre-chamber wall, displayed in picture (a) in Figure 4.34. Thus the calculated length of the spark is similar to the SGSP, but the area that is actually reached by the spark is much bigger. For the correct spark velocity, the distance of the sliding contact point on the pre-chamber wall needs to be added to the spark length, which is then divided by the period of time. The correction of the velocity is shown in graph (b) in Figure 4.34. The high velocity designs (V1, V2, V5 and V6) show a significant increase in spark velocity that can be connected to the increased gas flow velocity close to the pre-chamber wall.

While the external position of the electrode gap is beneficial for the spark deflection, its impact on the combustion is less advantageous. The start of the combustion close to the pre-chamber walls increases heat losses of the flame kernel, as approximately 180° of the flame propagation is blocked by the wall. The risk of preignition, due to overheating of the extended center electrodes, is also significantly higher.

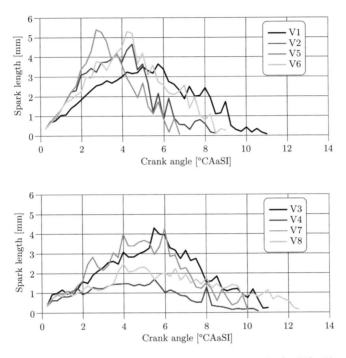

Figure 4.33: Spark deflection of the multi spark point spark plug (MSP). The orifice design with a higher deflection (V1, V2, V5, V6) are grouped in the upper graph

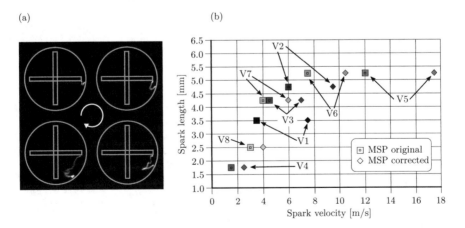

Figure 4.34: Example of the sliding ground contact point on the pre-chamber wall in (a). Correction of spark deflection velocity of the MSP spark plug for the eight orifice designs in (b) [129]

4.8.4 Comparison of the different spark plug designs

A beneficial air movement for the spark formation inside the pre-chamber can be achieved by different means. A given orifice design can cause different air movements in combination with different pre-chamber designs. In Figure 4.35 the maximal spark length and spark velocity of all tested combinations are displayed. The spark length is not corrected by the distance between the electrodes.

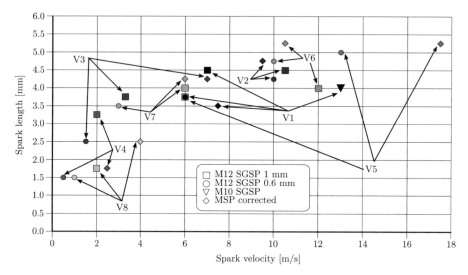

Figure 4.35: Summary of the tested configurations and the impact of the orifice designs in combination with the different pre-chamber shell designs regarding the spark deflection and velocity [129]

The spark velocity is calculated with Equation 3.2. The velocity of the MSP spark plug is corrected as in Figure 4.34. Regarding the orifice designs, the following conclusions can be drawn due to the direct comparison:

- The V1 design that is tested in all spark plug configurations, shows the highest spark velocity with the M10 spark plug. Even though the MSP spark plug has the same inner diameter of 7 mm and a more favorable electrode gap position, its spark deflection and velocity is smaller.

- The V2 design with a smaller hole diameter of 0.6 mm causes an increase in velocity and spark length, the achieved values are similar for all pre-chamber designs.

- An increase of the hole diameter to 1 mm (V3) reduces the spark deflection and velocity. Only in combination with the MSP spark plug, the values are close to the V1 design. However the results are still higher than the four hole design with a similar orifice surface (V8).

- An orifice design without hole offset (V4) hardly generates a spark deflection. However, a greater ED helps to achieve a spark length of 3.25 mm.

- The highest spark velocity is attained with an increase of the hole offset. The V5 design (1 mm hole offset) in combination with the electrode position and the small internal diameter of the MSP spark plug reaches the global maximum of 17.5 m/s. For the SGSP the 0.6 mm ED shows better results than the 1.0 mm.

- The V6 design with four instead of six holes shows similar results like the V2 design with its hole diameter reduction from 0.8 mm to 0.6 mm. The spark velocity is comparatively high with up to 12 m/s, the spark length reaches with 5.25 mm a global maximum.

- The combination of four holes with a bigger diameter of 1 mm (V7) has a similar orifice surface as the V1 design. The results of both designs are also comparable. However, the V7 M12 SGSP with 0.6 mm ED shows only half the spark velocity than the V1 design.

- The V8 design with four bigger holes of 1.2 mm diameter hardly creates a spark deflection. The results are similar to the V4 design without offset, apart from the observation that an increase in ED has no positive effect. The six hole design with the same orifice surface (V3) shows a higher spark deflection.

For the pre-chamber shell designs it can be summarized:

- A smaller ED results in longer sparks, when the air movement in the electrode gap is increased.

- A bigger ED can increase the spark length, when the air movement is little in the electrode gap.

- The inner diameter of the pre-chamber has a significant impact on the spark deflection. Smaller diameters result in higher deflection velocities of the spark.

- An electrode gap close to the cylindrical spark plug wall causes a sliding of the ground contact point. Even tough the spark length is not necessarily higher than with a central electrode gap, the spark can reach a remarkably larger area.

4.9 Pressure measurement inside the pre-chamber

4.9.1 Pressure inside the M12 pre-chamber measurement spark plug

The pressure inside the M12 pre-chamber spark plug of the IFKM research engine is measured with a cylinder pressure sensor as described in Figure 3.2. When the MCC pressure is subtracted from the pre-chamber pressure, the resulting differential pressure is giving an indication about the flow direction of the gas between both volumes. A positive flow means hereby a flow from pre-chamber towards MCC. Figure 4.36 shows the pressure in pre-chamber, MCC and the differential pressure throughout the engine cycle.

During the intake stroke ($-360\,°\mathrm{CA_{aTDCf}}$ to $-180\,°\mathrm{CA_{aTDCf}}$) the measured pressure in pre-chamber and MCC are equally at approximately 0.8 bar, which corresponds to the pressure in the intake system. The filling of the pre-chamber happens during the compression stroke where unburned mixture is compressed by the piston in the cylinder ($-180\,°\mathrm{CA_{aTDCf}}$ to $-9\,°\mathrm{CA_{aTDCf}}$). The pressure difference between both volumes increases hereby after $-60\,°\mathrm{CA_{aTDCf}}$, pushing fresh mixture into the pre-chamber via the orifices. Once the plasma between the spark plug electrodes inside the pre-chamber starts the combustion at $-15\,°\mathrm{CA_{aTDCf}}$, the pressure rises above the level of the MCC at $-9\,°\mathrm{CA_{aTDCf}}$. During these $6\,°\mathrm{CA}$, the compression of the piston compensates the pressure increase in the pre-chamber. At $-9\,°\mathrm{CA_{aTDCf}}$ the ejection of the jets begins and lasts for a duration of $8\,°\mathrm{CA}$. Shortly after the pressure in the pre-chamber reaches its peak ($-5\,°\mathrm{CA_{aTDCf}}$) the pressure rise in the MCC starts, indicating that the combustion in MCC has passed the early flame development stage. The pressure difference decreases, as the pre-chamber combustion comes to an end and the MCC pressure increases. At $-2.3\,°\mathrm{CA_{aTDCf}}$ the pressure is equal in both volumes. Since the combustion in the MCC continues, the rising cylinder pressure pushes gas into the pre-chamber from $-2.3\,°\mathrm{CA_{aTDCf}}$ to the crank angle with the maximum MCC pressure (α p_Cyl max) at $14\,°\mathrm{CA_{aTDCf}}$. After reaching its maximum (p_Cyl max), the pressure in the MCC decreases and causes the emptying of the pre-chamber until $90\,°\mathrm{CA_{aTDCf}}$. The end of the cycle ($90\,°\mathrm{CA_{aTDCf}}$ to $360\,°\mathrm{CA_{aTDCf}}$) shows no measurable pressure difference between both volumes.

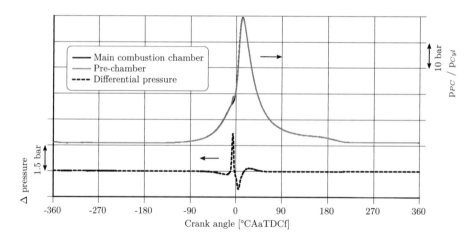

Figure 4.36: Pressure measurement inside the pre-chamber and the MCC. The differential pressure (Δ pressure) is additionally displayed

From a mechanical point of view, the negative pressure difference – from end of jet ejection to α p_Cyl max in the MCC – is critical for the pre-chamber cap. The heat of the combustion decreases the mechanical properties of the cap material. If the cap is not correctly designed, the pressure on its surface can cause a deformation. This is the reason why overheated caps show a concave cap deformation, despite that the highest differential pressure is from the inside towards the outside of the pre-chamber. In Figure 4.37 the measurement of a cap surface is displayed that was deformed during engine use. The center of the cap is pushed almost $0.4\,\mathrm{mm}$ to the inside, as the the shape of a new cap is flat. The evenness of the cap after engine testing can therefore be an indicator about the limits of the cap regarding its temperature and pressure withstanding geometry.

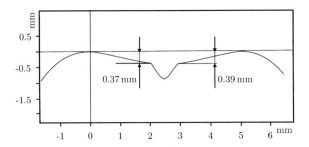

Figure 4.37: Contour measurement of a pre-chamber cap with a deformation of approximately 0.4 mm. The centering in the middle of the cap is for manufacturing reasons

During preignition events the combustion starts premature in the cycle. A positive pressure difference due to a jet ejection is hardly or not present (Figure 4.2). This can also lead to a concave pre-chamber cap deformation.

The pressure in the pre-chamber during its combustion has a direct impact on the hot gas jets that are ejected. As those jets are responsible for the ignition of the mixture in the MCC, the resulting cylinder pressure is also connected to the pre-chamber pressure. A data comparison of two OPs at 2000 rpm, 8 bar IMEP and a λ of 1 and 1.3 shows this dependency in Figure 4.38.

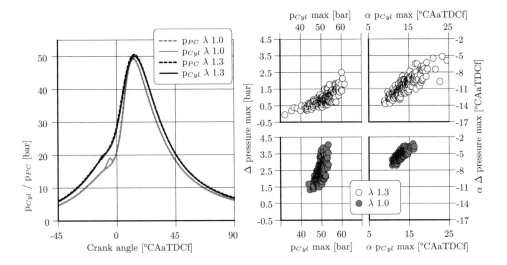

Figure 4.38: Comparison of maximal pressure difference between pre-chamber and main combustion chamber (Δ pressure max) and maximal cylinder pressure (p_{Cyl} max) and the crank angle position of both maximal pressures (α p_{Cyl} max and α Δ pressure max)

For the leaner OP the air flow is adjusted to the same fuel mass as at λ=1, thus the IMEP is increased to 8.3 bar due to the higher efficiency of the leaner combustion. The ignition timing is adjusted for a MFB50 at $8\,°CA_{aTDCf}$. The average of the 200 recorded cycles of both OPs shows that the pre-chamber does not work properly at the leaner OP. While the λ=1.0 OP has a distinct pressure bump after ignition, the λ=1.3 OP shows only a small increase. The cause therefore is that the ratio *pressure rise due to fuel energy in the pre-chamber* versus *pressure rise due to compression in the MCC* is less favorable with increased air in the MCC. One reason is that less fuel is trapped in the pre-chamber due to the earlier ignition timing. The other reason is that the combustion in the pre-chamber is slower and thus the pressure increases less rapidly. The throttle effect of the holes is therefore reduced.

As the lean operating pre-chamber is less efficient, the following combustion in the MCC suffers. XY-plots with max. p_{Cyl} and α p_{Cyl} max of pre-chamber and MCC for each cycle display the combustion correlation between both volumes in Figure 4.38.

The envelope of the reached maximal MCC pressures is significantly smaller in richer conditions (X-axis). The range of the maximal pre-chamber pressures (Y-axis) is comparable between both OPs, with an offset of approximately 1 bar between them.

While the lean OP has a strong connection between a high pre-chamber and a high resulting MCC pressure, the richer OP shows only a little dependency. The same behavior can be stated for the crank angle position of maximal pressure (α p_{Cyl} max).

The design of an adapted pre-chamber for the leaner OP would have an increased volume, to trap more fuel energy for the pre-chamber combustion. It would also be beneficial to reduce the orifice surface towards the MCC and therefore increase the throttling during the pre-chamber combustion.

4.9.2 Pressure inside the M14 pre-chamber insert

The operation range of the Renault engine – in terms of engine speed and load per cylinder – is challenging for the development of the pre-chamber spark plug. The mechanical stress that is imposed on the pre-chamber cap, due to the pressure difference between pre-chamber and MCC, together with the thermal load of a fuel mass flow of 16.67 kg/h per cylinder, burned with a stochiometric or leaner air-fuel ratio, make it necessary to know the boundary conditions of the combustion system.

The pressure measurement inside the pre-chamber is therefore realized with a cylinder pressure sensor, which is connected via a drilled channel in the cylinder head to the pre-chamber volume (Figure 3.3). The pre-chamber is designed as an insert that ports a spark plug. As a consequence, the pre-chamber volume can be varied by $125\,mm^3$ by changing the spark plug position in the insert.

The λ is varied by 0.1 points during the experiment with the bigger volume. The orifice design for the pre-chamber consists of six holes with a diameter of 1.2 mm, there is no offset of the holes to the spark plug axis. Engine speed for the OPs is set to 10500 rpm, the compared data are average values of 300 cycles. Figure 4.39 shows the pressure difference between the pre-chamber and the MCC, together with the pressures in both volumes. A negative pressure means hereby a flow from the MCC into the pre-chamber, a positive pressure difference provokes a flow from the pre-chamber into the MCC. The 50 bar pressure difference during the hot gas ejection is relatively high, but can be found on the same level in gas engines [56, 96, 99]. A variation of the pre-chamber volume at a fixed air-fuel ratio requires different ignition timings to reach the knock limit of the engine.

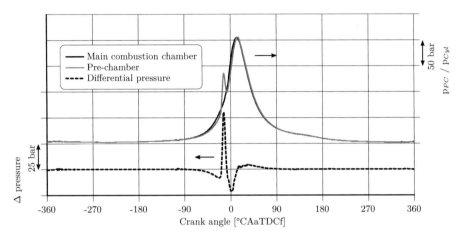

Figure 4.39: Pressure measurement in main combustion chamber and pre-chamber together with the resulting differential pressure

A smaller volume demands an earlier ignition timing of 1.5 °CA and results into an earlier pressure rise in the pre-chamber. Graph (a) in Figure 4.40 shows the pressure measurement for both pre-chamber volumes at the same air-fuel ratio. Besides the pressure curve in MCC and pre-chamber the pressure difference between both volumes and the current of the ignition coil is displayed. Before the ignition both volumes are filled during the compression stroke. The comparison of the differential pressure shows a more negative value for the bigger pre-chamber volume.

This means that the density of the mixture in the larger volume is lower than in the smaller pre-chamber at the same crank angle. However, the later start of combustion of the bigger volume leaves more time for its filling.

Together with the pressure rise in the MCC of the continuing compression the maximal pressure during the pre-chamber combustion is 15 bar higher with the bigger pre-chamber. The engine shows the same maximal cylinder pressure at the same crank angle position and the same BSFC for both pre-chambers. This implies that the later jet ejection of the bigger pre-chamber is compensated by the higher pressure difference during ejection. The resulting MCC combustion is faster in the beginning to reach the same pressure curve characteristics.

If air is added and the engine is operated under leaner conditions with the same fuel mass flow and pre-chamber volume, more ignition advance is applied to reach the knock limit of the engine (graph (b) in Figure 4.40). An interesting observation is the position of the maximal pressure in the MCC, which is 2 °CA earlier in the cycle. The knock limit for the diluted OP is therefore at an earlier °CA – despite its slower burn velocity – than for the richer mixture. The explanation for this behavior is the decrease of the knock tendency due to a colder pre-chamber ignition system. This theory is confirmed by a test at richer conditions where preignition occurs.

The maximal pressure is increased by 20 bar and the efficiency of the engine is improved by $-5\,\mathrm{g/kWh}$ BSFC for the leaner OP. An observation is the smaller pressure bump for the more diluted OP during the hot gas ejection. As fuel measurements inside the pre-chamber reveal, the fuel amount is equal for both OPs at the same crank angle [128]. However, as the ignition is earlier, less energy is trapped.

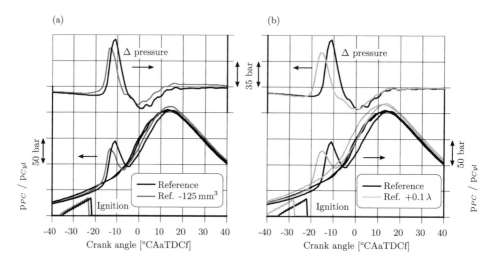

Figure 4.40: Pressure measurement with different pre-chamber volumes (a) and air-fuel ratios (b). Together with the pressure in the different volumes the differential pressure and the ignition current signal is displayed

A retreat of ignition timing results in a later pressure peak in the pre-chamber. The offset between the ignition timing and the $\alpha\ p_{Cyl}$ max of the pre-chamber stays hereby constant at $14\,^\circ$CA for the rich and $16\,^\circ$CA for the lean OPs. Graph (a) in Figure 4.41 shows the evolution of the pressure peak for the three variants during the ignition timing sweep. The increase of λ by 0.1 with the bigger pre-chamber results in a about 10 bar lower pressure difference between pre-chamber and MCC (Δ pressure max) at the same crank angle with the maximal pressure difference between pre-chamber and MCC ($\alpha\ \Delta$ pressure max).

It is remarkable that Δ pressure max with a small volume pre-chamber is almost constant during the ignition retreat. The increase of air-fuel mixture that is additionally flowing into the pre-chamber, due to the later ignition in the engine cycle, is not transformed into a higher peak-pressure as with the bigger volume. In other words, this means that the smaller volume keeps a constant pressure difference at earlier ignition, while the pressure difference for the bigger volume decreases.

The maximal pressure gradient during the pre-chamber combustion (Δ pressure gradient max) is displayed in graph (b) of Figure 4.41 at its corresponding $^\circ$CA ($\alpha\ \Delta$ pressure grad. max). The bigger pre-chamber shows the fastest combustion and therefore the highest pressure gradient. A dilution of the mixture reduces significantly the speed of the pressure rise, as does the smaller volume. While the gradient decreases for later ignition timings for the richer mixture, the lean OP shows no deviation. The ejection duration is similar for all pre-chambers at $10\,°\text{CA}$ and increases with later ignition advance.

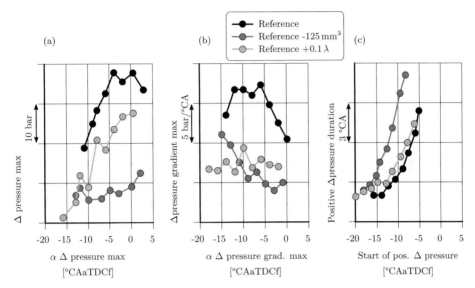

Figure 4.41: Pre-chamber pressure measurement during ignition sweep. Maximal pressure difference and its position (a), maximal gradient of the differential pressure and its position (b) and duration of the positive pressure at its start (c)

Graph (c) in Figure 4.41 displays the ejection duration (positive Δpressure duration) at its beginning (start of positive Δ pressure). A look at the crank angle resolved differential pressure gives an indication, why the pressure difference of the small volume pre-chamber is less dependent on the ignition timing than the bigger volume (graph (a) in Figure 4.42). The area of the negative pressure – filling of the pre-chamber – is larger between the curves for $2\,°\text{CA}$ ignition retreat with the bigger volume. The orifices that connect the pre-chamber with the MCC throttle the gas flow into the pre-chamber during the compression stroke. A bigger volume pre-chamber is more impacted by this effect, as more mass needs to flow through the nozzles to reach the same pressure level. As the volume difference between the two pre-chambers is negligibly small compared to the volume of the MCC, the pressure in the MCC during compression can be considered as equal for both OPs. Therefore, an increased negative pressure difference means a less dense media in the pre-chamber. This can be seen in graph (a) in Figure 4.42 where the pressure difference of the bigger pre-chamber is almost two times higher than for the smaller volume.

The ideal gas law is used to estimate the mass (m_{PC}) in both pre-chambers before ignition. Both pre-chamber volumes are compared at the same air-fuel ratio, it is assumed that the mixture preparation is homogeneous in the engine and therefore the individual gas constant (R) is the same for both OPs. The temperature of the gas (T_{PC}) is considered as equal during compression, differences in the compression ratio due to the larger pre-chamber volume are therefore not taken into account. Thus the mass in the pre-chamber is only dependent on the pressure (p_{PC}) and the pre-chamber volume (V_{PC}).

$$p_{PC} \cdot V_{PC} = m_{PC} \cdot R \cdot T_{PC} \qquad (4.2)$$

With this estimation, the known pre-chamber volume (V_{PC}) and the measurement of the pressure inside the pre-chamber (p_{PC}), the trapped mixture mass (m_{PC}) in the pre-chambers can be calculated:.

$$m_{PC} = p_{PC} \cdot V_{PC} \qquad (4.3)$$

In graph (b) in Figure 4.42 the calculation with equation 4.3 is displayed. During the compression and before ignition the smaller pre-chamber has an offset of 6 °CA before it is filled with the same amount of mixture as the bigger pre-chamber. However, even with the lower density in the pre-chamber the mass in the bigger volume is still superior to the small volume pre-chamber at the same °CA.

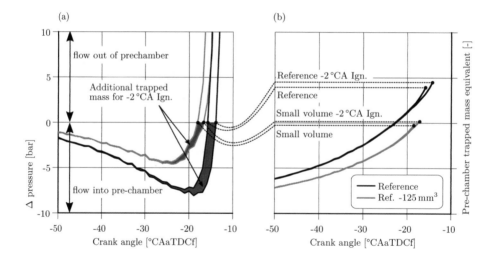

Figure 4.42: Differential pressure for a ignition retreat of 2 °CA for the bigger and smaller pre-chamber at the same air-fuel ratio in (a), the area represents the additional trapped mass in the pre-chamber. The calculated trapped mass in the pre-chamber is displayed in (b) for both pre-chambers

The combustion stability in the pre-chamber can be connected to the variation of the maximal pre-chamber pressure during the hot jet ejection. A stable and repeatable combustion results in a small coefficient of variation of the maximal differential pressure between pre-chamber and MCC (COV Δ pressure max).
The impact of the air-fuel ratio and the pre-chamber volume on the combustion stability in MCC (COV$_{IMEP}$ in graph (a)) together with the variations of the differential pressure

between MCC and pre-chamber (COV Δ pressure max in graph (b)) are displayed in Figure 4.43. Both COVs are plotted over the BSFC for an ignition sweep. The best performance is achieved by the lean operated bigger volume pre-chamber, despite a slightly higher COV_{IMEP}. All pre-chambers show a linear connection between combustion stability and engine efficiency for later ignition timings.

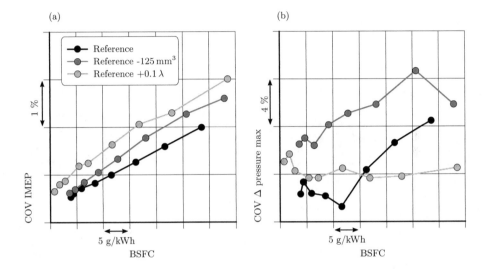

Figure 4.43: Combustion stability (COV_{IMEP}) of the engine (a) and stability of the maximal pressure difference (COV Δ pressure max) between pre-chamber and MCC (b)

The combustion in the smaller pre-chamber volume is less stable with an COV Δ pressure max increase of 5 % for a later ignition. The bigger volume shows the best max. Δp stability and as well an increase of 5 %. The leaner operated pre-chamber stays at a stable value for all ignition timings, with a COV Δ pressure max in between the values of the pre-chambers in rich conditions. The best performance is achieved with the bigger volume in lean conditions. There is no performance difference between both volumes at richer conditions.

4.10 Optical measurement inside the combustion chamber

A look inside the combustion chamber is enabled by the different accesses of the IFKM single cylinder engine. An endoscopic measurement of pre-chamber ignition systems at the IFKM is described by Disch et al. and includes the chemiluminescence investigations and a complex post treatment of the acquired data for a single jet [35]. The pre-chamber orifice design has therefore a hole in a distance of minimal 120° towards the other holes, to exclude jet interference. For the pre-chamber design in the current work, this restriction was not granted. The endoscopic measurement is therefore analyzed visually, as the information output is judged less relevant for the analyzed hypothesis compared to the other techniques that are presented in this work[7]. For the optical comparison between the spark ignition by a standard spark plug and the jet ignition by the pre-chamber a LaVision HSS6 camera together with a Storz 6.5 mm endoscope and a light source are used. The frequency of the recorded pictures is 12 kHz which corresponds to 1 picture/°CA. The OP is at 2000 rpm and 8 bar IMEP at λ=1. The standard spark plug has an ED of 1 mm. The pre-chamber spark plug has a six hole design with a hole diameter of 0.8 mm, a volume of 350 mm^3 and an ED of 0.2 mm. The MFB50 for the compared cycle is set to 7 °CA$_{\mathrm{aTDCf}}$.

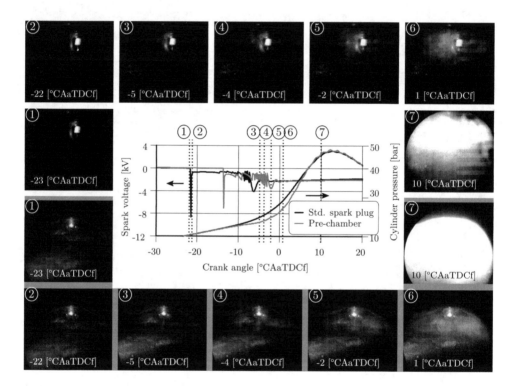

Figure 4.44: Endoscopic comparison of pre-chamber and standard spark plug. Spark voltage and cylinder pressure is displayed together with seven endoscopic pictures at characteristic crank angle positions

Figure 4.44 shows pictures of the combustion at different representative crank angles, beginning with the spark breakthrough of the standard spark plug at -22 °CA$_{\mathrm{aTDCf}}$. The air movement between the electrodes is little, this can be deduced by the progress of the spark voltage, which stays constant. In the cycle of the standard spark plug 10 % of the fuel is burned (MFB10) at -4 °CA$_{\mathrm{aTDCf}}$, the appearance of the spark plug in the

---

[7]Focus in this work was the optical measurement of HC and CO_2 inside the pre-chamber by an IR sensor and the spark deflection inside the pre-chamber

corresponding picture is therefore blurry, due to density change of the air between the focused object and the endoscope. During the previous flame development angle of 17 °CA the camera pictures show now visual combustion. In the cycle with the pre-chamber the early jet ejection can be discovered at -5 °CA$_{\mathrm{aTDCf}}$, which is 9 °CA after ignition and only 4 °CA before MFB10. Shortly after TDC the combustion propagation of the pre-chamber cycle shows a greater area in the combustion chamber than the standard spark plug. The reason therefore are the multiple ignition spots of the jets and the resulting increased flame front, whereas the standard spark plug has a spherical flame propagation originating from the spark plug. Additional measurement of the pre-chamber pressure together with the images of the high speed camera combines the visual hot gas jet ejection with the pressure difference between pre-chamber and MCC.

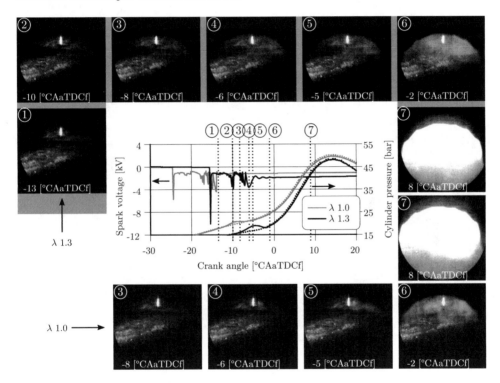

Figure 4.45: Optical measurement with pre-chamber at λ=1.0 and λ=1.3. Pictures at the same crank angle show the different evolution of the flame in the combustion chamber. The pressures in pre-chamber and main combustion chamber are displayed together with the spark voltage

Figure 4.45 shows the acquired data for a cycle at λ=1 and λ=1.3. The pre-chamber pressure measurement of the lean OP reveals a deficit for the pre-chamber (Figure 4.38). The high speed camera acquisition can therefore be of interest, to visualize the difference between the combustion initiation of the jets for both OPs. Pictures of the lean cycle show less frank jets during the early ejection, as a lower differential pressure pushes the gas into a denser mixture in the MCC. The benefit of a deeper jet penetration and therefore faster start of combustion can be seen by comparing the pressure rise in the MCC after the jet ejection in both cycles. The λ=1.3 cycle has a smaller pressure gradient in addition to a larger delay between jet ejection and MCC pressure rise. The visibility of the jets are directly connected to the overpressure in the pre-chamber, cycles with a higher pressure difference result in more pronounced jets.

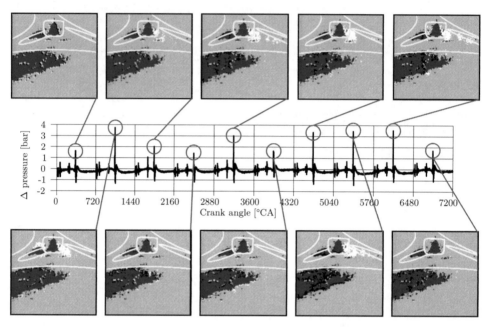

Figure 4.46: Optical measurement of cyclic variations at $-5\,°$CAaTDCf for cycles together with the pressure difference between pre-chamber and main combustion chamber

The cyclic variations of the pre-chamber differential pressure can therefore be found in the area of the visible combustion in the recorded images. Figure 4.46 shows the jet ejection at $-5\,°$CA$_{\text{aTDCf}}$ for ten consecutive engine cycles, the black area represents permanent reflections in the combustion chamber, the visible combustion is displayed in white. For cycles with a high peak pressure the picture shows more ejected gas in the MCC.

4.11 Determination of unburned fuel (hydrocarbons) inside the pre-chamber

The mixture in the passive pre-chamber depends on fuel that is injected outside of its volume into the MCC or before the inlet valves. The air-fuel mixture that reaches the ignition point inside the pre-chamber is brought there by the pressure difference between pre-chamber and MCC. The amount of energy that contributes to the combustion in the pre-chamber is proportional to the amount of fuel inside its volume. A measurement of the mixture composition inside the pre-chamber can therefore be helpful to develop different injection strategies or engine OPs to favor this fuel transport.

The local measurement of HC inside the pre-chamber is performed with the sensor spark plug described in Section 3.3.3. Fuel density is measured with an IR sensor during fired and motored OPs. The comparison between cold and hot environment gives an indication about the possible use of the sensor in a motored engine with the air and fuel amount of a high load OP. Conditions that are not accepted by the sensor and could cause preignition events due to overheating of the sensor mirror that protrudes into the combustion chamber. The sensor limits the engine utilization to a maximal IMEP of 8 bar and a maximal pressure of 200 bar (Table 3.2). A fired force inducted engine reaches these limits quite early, therefore mainly partial load measurements can be performed. However, the conditions of temperature and pressure in a motored engine without the combustion heat are far lower. Thus when fuel is injected into the motored engine, the most important unknown is the impact of fuel vaporization effects due to a colder combustion chamber. During the fired OPs, the engine is operated at 2000 rpm with IMEPs up to 8 bar. Fuel delivery into the engine is held constantly during the air-fuel ratio variation. The air-fuel ratio is adjusted via the air in the cylinder by the boost pressure.

The experiment is conducted with direct and indirect fuel injection. For the sensor calibration the indirect injection is used, as the external mixture preparation outside the cylinder results in a homogeneous air-fuel distribution.

4.11.1 Air-fuel equivalence ratio for fired and motored engine with direct and indirect fuel injection

The comparison in Figure 4.47 shows the air-fuel ratio inside the pre-chamber before the earliest applied ignition timing at -35 °CA$_{aTDCf}$ and before the earliest start of combustion inside the pre-chamber at -32 °CA$_{aTDCf}$. The applied ignition timing varies, as the vised MFB50 is set to 10 °CA$_{aTDCf}$ for all air-fuel ratios. The local measurement is compared to the values that are recorded by the LSU4.9 Lambda sensor in the exhaust. During the motored OPs, the ignition is cut off and the fuel injection is continued. Besides the mean value, the standard deviation of the air-fuel ratio measurement is also displayed in the graph. A larger bar means hereby a bigger range of measured values, this occurs especially during the OPs with direct injection. A smaller deviation can be observed in the motored engine. Especially the motored indirect OPs show very representative values. The comparison of the left and right graph in Figure 4.47 shows the rapid decrease of the air-fuel equivalence ratio in the pre-chamber between -35 °CA$_{aTDCf}$ and -25 °CA$_{aTDCf}$. While the motored OPs show less evolution (0.5 λ point), the values in the fired engine decrease significantly (1 λ point). The inhomogeneous mixture field in the MCC, with its central mounted direct injector, results in a richer mixture in the center of the MCC around the pre-chamber orifices. Therefore, the local λ is lower with direct fuel injection than with indirect fuel injection.

Fired OPs for direct and indirect injection have almost a constant offset between their average air-fuel ratios. Values for the motored injection are close to the fired points.

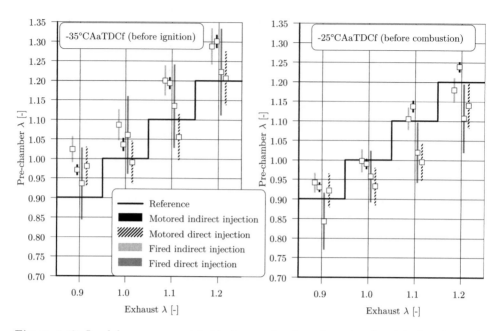

Figure 4.47: Lambda measurement inside the pre-chamber of a motored and fired engine with direct and indirect fuel injection compared to the exhaust lambda measurement (4 bar IMEP, 2000 rpm) [128]

Not surprisingly, the local λ in the pre-chamber follows for all OPs the trend of the exhaust measurement.

It may therefore be expected that the measurement of motored high load engine OPs allows a quantification of different strategies, regarding the fuel supply into the pre-chamber, and that the impact by the hot surfaces in the combustion chamber does not change the trend of the results.

4.11.2 Crank angle resolved lambda measurement

The crank angle resolved λ values show the evolution of the local HC density in the pre-chamber during the engine cycle. For the motored and fired OP the injection timing and duration is held constant. The injection timing for the direct injection is set to -300 °CA$_{aTDCf}$. The air is adjusted for an exhaust air-fuel equivalence ratio of 1.2. Figure 4.48 shows that the air-fuel ratio in the pre-chamber for the motored engine has its minimum slightly after TDC. The cause for this offset is a slower decrease of the fuel density inside the pre-chamber after TDC due to fuel that is stocked. Measurements in the MCC of a motored engine show the same effect [61], König et al. supposes an outgassing effect from the combustion chamber crevices [69]. For the fired and motored OPs, the direct injection shows a later decrease than the indirect injection. The offset of about 10 °CA is a consequence of the reduced time for the mixture preparation in the cylinder of the direct injection.

The fired cycle in Figure 4.48 displays the heat release rate of the combustion in the MCC. The sensor measures a lower maximal λ value with the direct injection before ignition.

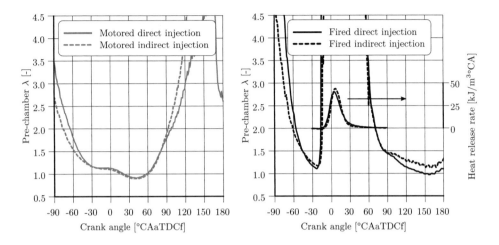

Figure 4.48: Crank angle resolved air-fuel ratio comparison of the motored and fired engine with indirect and direct fuel injection (4 bar IMEP, 2000 rpm) [128]

For both fired OPs, the MFB50 is held constant at 10 °CA$_{\text{aTDCf}}$, the applied ignition timing differs slightly between direct (-34 °CA$_{\text{aTDCf}}$) and indirect injection (-33.5 °CA$_{\text{aTDCf}}$). A later disappearance of the fuel – and consequently very high λ values – indicates the begin of the combustion in the pre-chamber. The heat release rate of the combustion in the MCC shows a faster combustion with the homogeneous fuel distribution.

An unexpected discovery is the measurement of HC in the pre-chamber after combustion. The measurement suggests that during the early combustion phase in the MCC, the pressure differences cause a refilling of the pre-chamber with unburned gas. Another explanation could be the quenching of the flame on the pre-chamber walls. The direct comparison of equal OPs with fuel density and pressure measurement in the pre-chamber are displayed in Figure 4.49. After the hot gas ejection (-10 °CA$_{\text{aTDCf}}$ to -2.5 °CA$_{\text{aTDCf}}$) the pressure rise in MCC pushes gas into the pre-chamber (-2.5 °CA$_{\text{aTDCf}}$ to 15 °CA$_{\text{aTDCf}}$). The IR sensor measures a presence of fuel in the pre-chamber beginning at 45 °CA$_{\text{aTDCf}}$. A rational expectation is that all available fuel is burned during the combustion process, and the absorption signal is therefore close to zero [16, 60].

However, the presence of fuel after combustion was also discovered by Koenig and Hall or Grosch et al., both publications hold outgassing cavities responsible for the HC [45, 69]. Different theories explain the presence of fuel after the combustion. First cause could be the quenching of the flame on the pre-chamber walls, resulting in measurable HC emissions. Another theory is is the quenching of the flame in the pre-chamber holes of the inflowing mixture during the main chamber combustion. An additional hypothesis is the presence of unburned mixture in the center of the MCC around the pre-chamber holes. While the hot gas jets ignite the flame further outside the MCC, unburned mixture flows into the pre-chamber at the early MCC combustion. Optical measurement with transparent pistons show the presence of unburned fuel close to the holes almost until the end of combustion [8, 118].

Nevertheless, as the sensor is not calibrated in the hot environment after the combustion, the absolute value of the measurement is invalid.

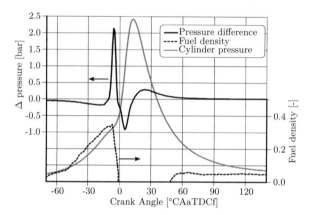

Figure 4.49: Crank angle resolved fuel density inside the pre-chamber together with the shape of the cylinder pressure and the pressure difference between pre-chamber and main combustion chamber [128]

4.11.3 Impact of injection timing and global air-fuel ratio on the fuel delivery into the pre-chamber

In an engine with direct fuel injection the mixture preparation is highly reliant on the injection timing. As the air-fuel ratio in the passive pre-chamber is dependent on the mixture that enters its volume via the MCC, a significant change of the evolution of the local fuel content in the pre-chamber can be seen.

Figure 4.50 displays a SOI sweep from -360 °CA$_{aTDCf}$ to -240 °CA$_{aTDCf}$ for OPs with 4 bar IMEP. The ignition timing is adjusted for a MFB50 at 10 °CA$_{aTDCf}$. An earlier decrease of the air-fuel ratio inside the pre-chamber is preferable, as this is a sign for a more complete mixture of fuel and air, as therefore more time is available. Additionally, the burnable air-fuel ratio ($\lambda=0.7$ to $\lambda=1.2$ [55]) is also reached prior. The earliest presence of fuel is measured with a SOI at -300 °CA$_{aTDCf}$ and -290 °CA$_{aTDCf}$, resulting in good combustion results for the 4 bar IMEP. Later SOIs than -280 °CA$_{aTDCf}$ cause a reduced and unstable fuel delivery into the pre-chamber, by which the combustion stability is significantly decreased.

The comparison of the COV$_{IMEP}$ and the MFB50 of a pre-chamber and a standard spark plug shows the unstable combustion for late SOIs in Figure 4.51. For the experiment the ignition timing is held constant at -22 °CA$_{aTDCf}$ for the standard spark plug and -18.5 °CA$_{aTDCf}$ for the pre-chamber.

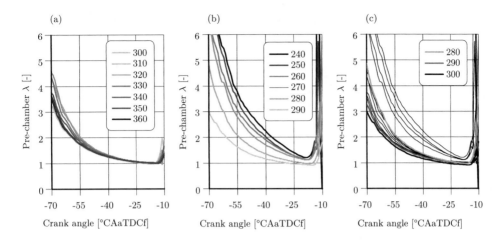

Figure 4.50: Crank angle resolved fuel density for a SOI sweep between -360 °CA$_{aTDCf}$ and -240 °CA$_{aTDCf}$. Early SOI in (a), later SOI in (b), best working SOIs in (c) (4 bar IMEP, 2000 rpm)

Remarkable are the two OPs with SOI at -290 °CA$_{aTDCf}$ and -280 °CA$_{aTDCf}$, where the pre-chamber shows a faster combustion than with an earlier injection. This is against the trend of the standard spark plug, with the latest MFB50 for these two OPs.

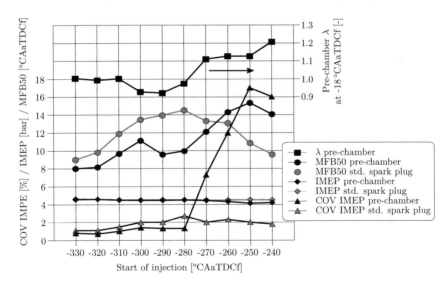

Figure 4.51: Combustion results during SOI sweep with constant ignition timing for standard spark plug and pre-chamber with local fuel measurement inside the pre-chamber (4 bar IMEP, 2000 rpm)

It can therefore be assumed that the faster combustion is not caused by the mixture in the MCC – as this would also be beneficial for the standard spark plug – but by a change of the air-fuel ratio inside the pre-chamber. The local fuel measurement confirms this theory, as the measured mixture composition is comparably rich.

Combustion and air-fuel ratio measurement results of the 8 bar IMEP SOI sweep are compared in Figure 4.52.

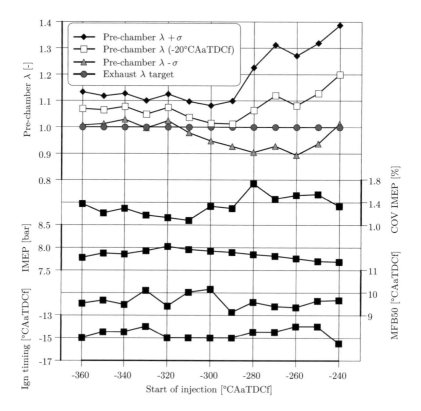

Figure 4.52: Combustion results and fuel measurement inside the pre-chamber at $-20\,°\mathrm{CAaTDCf}$ during a SOI sweep at 8 bar IMEP and 2000 rpm [128]

The displayed λ is measured at $-20\ °\mathrm{CA_{aTDCf}}$, the standard deviation is added/subtracted to the mean value to illustrate the variations of the mixture composition. The ignition timing is adjusted for a MFB50 at $10\ °\mathrm{CA_{aTDCf}}$. A retreat of the injection timings from $-360\ °\mathrm{CA_{aTDCf}}$ shows a trend for an enrichment of the pre-chamber. Most fuel is measured at $-300\ °\mathrm{CA_{aTDCf}}$ and $-290\ °\mathrm{CA_{aTDCf}}$, a later injection increases the average λ value and its standard deviation.

The combustion stability in the MCC is directly impacted by the combustion and the mixture preparation in pre-chamber and MCC, it shows a higher $\mathrm{COV_{IMEP}}$, if the fuel is injected later in the cycle.

The best engine performance is achieved with an injection timing at $-320\,°\mathrm{CA_{aTDCf}}$, with an IMEP of 8 bar. As the air-fuel ratio in the pre-chamber is comparable to the earlier SOIs, the performance is more connected to the beneficial mixture formation in the MCC.

The evolution of the pre-chamber air-fuel ratio during the engine cycle points out the impact of the SOI on the available energy in the pre-chamber. Some degree in crank angle before TDCf significantly change the fuel amount in the pre-chamber (Figure 4.49). The air-fuel ratio has the biggest impact on the ignition timing of an engine, with a demand of an earlier ignition for leaner mixtures due to the slower combustion. During the experiment, the injected amount of fuel is held constant and the engine is operated with a higher boost pressure, to dilute the mixture. The MFB50 is set to $10\,°\mathrm{CA_{aTDCf}}$ for all OPs. The earlier ignition timings for the leaner OPs can be seen in Figure 4.53 with an earlier disappearance of fuel before TDCf.

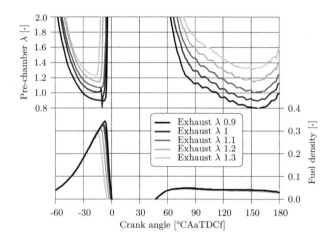

Figure 4.53: Fuel density and pre-chamber λ measurement during air-fuel ratio variation with different targeted exhaust λ at 8 bar IMEP for a constant MFB50 [128]

The λ value in the pre-chamber correlates with the exhaust measurement before the combustion. The measured fuel density shows that there is hardly a difference of the amount of fuel that reaches the sensor before ignition for the different OPs. The experiments reveal that the fuel support of the pre-chamber depends on the injected amount of fuel and its timing. The air in the combustion chamber has no impact. Thus, the energy in the pre-chamber does not depend on the global air-fuel ratio of the engine but on the injected fuel mass. However, the trapped mass in the pre-chamber is reduced with leaner OPs, as an earlier ignition timing is applied for the same MFB50.

The pressure measurement in Figure 4.49 shows that the filling of the pre-chamber happens between $-60\,°\mathrm{CA_{aTDCf}}$ and the ignition. A possible enrichment of the pre-chamber would therefore request a richer mixture outside the pre-chamber during this period. To investigate a local fuel support at late crank angle, a post injection is added to the main injection at $-300\,°\mathrm{CA_{aTDCf}}$. The main injection duration is therefore reduced from 1.5 ms to 1.3 ms and a second injection is added with 0.25 ms. The SOI of this post injection is varied from $-115\,°CA_{aTDCf}$ to $-35\,°CA_{aTDCf}$. The total injected fuel mass is equal for all OPs and result with the adjusted boost pressure in a λ of 1.4. The picture in Figure 4.54 shows the end of a post injection event. Event though visibly fuel is injected, there is no difference in the measured fuel density. The air-fuel ratio inside the pre-chamber has the same level and gradient for all OPs, the standard deviation is not increased by the post injection. While the conditions on the IR sensor position in the pre-chamber do not differ between a single and a multi injection, the combustion in the MCC changes. A second fuel supply impacts the IMEP of the engine, resulting in a performance loss of up to 1.6%. Cyclic variations of the main combustion increase with later injections to a COV_{IMEP} of 3.3%. Therefore, a local enrichment close to the pre-chamber does not stabilize the lean combustion.

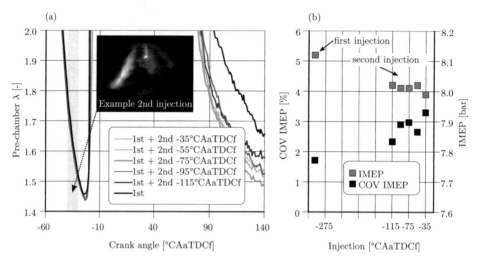

Figure 4.54: Comparison of single and multi fuel injection. A picture of a post injection event and the λ measurement inside the pre-chamber in (a) and the IMEP and combustion stability in (b)

It can be assumed that the time from the post injection to ignition is not sufficient to vaporize the fuel properly. Thus, no additional fuel is brought to the electrode position before plasma formation and the combustion in the MCC shows glowing of soot due to the poor mixture preparation.

However, after combustion more fuel is measured inside the pre-chamber when a post injection is performed.

4.12 Determination of burned fuel (carbon dioxide) inside the pre-chamber

The measurement of the burned gas fraction in the pre-chamber helps to understand the mixture composition, especially at the ignition begin in the pre-chamber. As the burned gas residuals have a major impact on cyclic variations and could cause misfires in the pre-chamber, knowledge about their content in the pre-chamber is of great interest.

By the use of the IR sensor pre-chamber spark plug, the local CO_2 percentage in the pre-chamber can be determined. Hereby, the temperature in the combustion chamber is calculated with the isentropic equation for ideal gas and is based on two temperatures (Section 3.3.3). The mean calculated temperature of a reference 8 bar IMEP $\lambda=1$ point is 374 °C. The measured temperature at the absorption volume is with 316 °C (Table 3.2) slightly lower. The difference can be explained by the inertia of the steel body of the measuring sensor (Figure 3.4) compared to the calculated gas temperature.

The impact of mixture dilution on the residuals in the pre-chamber spark plug is investigated for 4 bar and 8 bar IMEP OPs. The highest percentage of CO_2 can be measured at $\lambda=1$. With richer mixture carbon monoxide is formed due to the lack of oxygen. For leaner mixtures the surplus of oxygen can be measured in the burned gas.

The measurement with the IR sensor shows a decrease of the CO_2 content inside the pre-chamber after combustion. This behavior is surprising, since crank angle resolved measurements of other works show a constant CO_2 level after combustion [29, 44, 88]. Figure 4.55 shows the average CO_2 measurement of 200 cycles with the IR sensor together with the cylinder pressure for an OP at 4 bar IMEP (graph a) and 8 bar IMEP (graph b). The decreasing percentage of CO_2 at the sensor position supports the hypothesis of unburden mixture that enters the pre-chamber during the pressure rise in the MCC. The outflowing gas – during the expansion stroke – then increases the percentage of the fresh mixture in the pre-chamber. However, it could also show a remixing of the quenched air-fuel mixture on the pre-chamber wall with the rest of the gas at the sensor position.

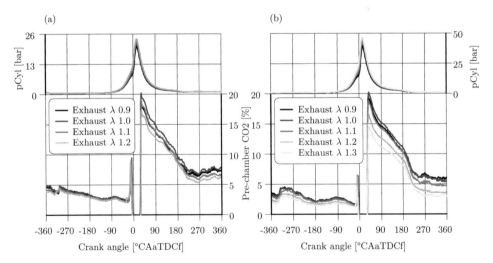

Figure 4.55: Cylinder pressure (pCyl) and CO2 measurement inside the pre-chamber during a air-fuel-ratio sweep for 4 bar (a) and 8 bar IMEP (b).

The measurement is contrary to simulation results, where the pre-chamber is filled with 100 % residual gas at -360 °CA$_{aTDCf}$, and the value decreases with filling of the pre-chamber with fresh mixture during the compression stroke [19, 65].

For both load points the CO_2 content decreases to about 10 % CO_2 before the exhaust valves open. The measured values during the compression start at 4 % and decrease to about 2 %. Even though the percentage might differ due to uncertainties in regards of the gas temperature, the curve course of the CO_2 stays correct. Concerns about burned gas residuals in the pre-chamber as a cause for misfires are therefore less relevant than the general gas movement between the electrodes in the pre-chamber.

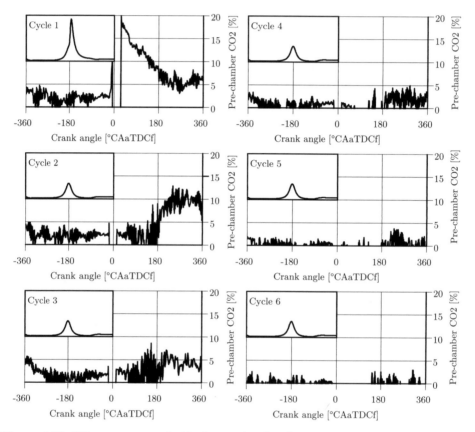

Figure 4.56: CO_2 measurement inside the pre-chamber during ignition cutoff. Cycle 1 is the fired cycle and cyles 2 to 6 are the consecutive cycles without combustion. The corresponding cylinder pressure is indicated in the top left of each graph (4 bar IMEP, λ=1 , 2000 rpm)

In Figure 4.56 six consecutive engine cycles are displayed with an ignition cut-off after the first cycle. The CO_2 content drops therefore during the expansion stroke for the first motored cycle. After opening of the exhaust valves, the backflowing burned gas residuals can be measured by the sensor. With ongoing cycles the CO_2 percentage decreases continuously.

4.13 Exhaust gas analysis with pre-chamber combustion

Exhaust emissions of the ICE are not restricted in the racing application field that is focused by the development. However, being directly connected to engine efficiency, they are of interest.

4.13.1 Measurement of hydrocarbon and nitrogen oxide emissions

The following comparison between the M12 pressure measurement pre-chamber spark plug (Figure 3.3) and a standard spark plug, focuses on HC and NO_x emissions of both ignition systems. The origin of HC emission is mainly unburned fuel, hence energy that is not transformed into power. A reduction to its maximum is the objective, even when in turbocharged engines an oxidation between exhaust valves and turbine still generates usable heat. NO_x emission are a sign of a fast combustion at hot temperatures. At a given air-fuel ratio, nitrogen oxides increase with engine efficiency [55]. The impact of pre-chamber designs and operation strategies on HC and NO_x emissions is investigated by Baumgartner et al. and compared to a standard spark plug. For the investigated small displacement engine of $500\,cm^3$, no significant emission difference between both ignition systems was found [13]. With an active pre-chamber Attard found for a given air-fuel ratio the same NO_x and lower HC emissions compared to a standard spark plug until the lean limit of the latter [7].

To compare exhaust emissions in the present work, OPs at 2000 rpm with 5 bar and 8 bar IMEP are chosen. The ignition timing is not set to the knock limit of the engine but to a MFB50 around $15\,°CA_{aTDCf}$. Figure 4.57 shows the combustion results (MFB10, MFB50, MFB90) together with NO_x and HC emissions during a SOI sweep. An OP with intake port injection and pre-chamber is added to each graph on the left side, to evaluate the impact of the mixture preparation of the direct fuel injection. The ignition timing is held constant during the experiment. The standard spark plug shows a stable combustion behavior during the SOI sweep. The combustion with a pre-chamber is far more dependent on the injection timing, as the mixture in the pre-chamber varies significantly. This statement is based on Figure 4.53 in Section 4.11 where the measured fuel density in the pre-chamber is displayed during a SOI variation at 8 bar IMEP.

An earlier injection timing than $-310\,°CA_{aTDCf}$ reduces the combustion duration for both plugs. The increased spray interaction with the hot piston closer to TDC can be determined as beneficial to evaporate the liquid fuel and to improve its homogenization with the air in the cylinder. The result is a faster and earlier combustion in the engine cycle. The emission of NO_x increases due to the faster combustion. Even though the intake valves are closed during injection, the HC emission increases slightly due to the fuel-piston interaction.

If the fuel is injected at later °CAs, the emission of HC increases as less time is available for the mixture preparation and the combustion is therefore less complete. The NO_x in the exhaust gas shows the associated decrease [54].

The pre-chamber ignition results in an unstable combustion with increased HC emissions for SOIs after $-280\,°CA_{aTDCf}$. The deterioration of the energy transformation can be connected to the unsteady mixture conditions inside the pre-chamber [128]. Therefore, the pre-chamber combustion cannot provide the repeatable hot gas jets to ignite the mixture in the MCC.

Even though the pre-chamber has a significant faster combustion at each OP, the NO_x emission is not significantly higher than with the standard spark plug, the MFB50 is at the same °CA as with the standard spark plug.

The strong correlation between the mixture conditions inside the pre-chamber and the combustion in the MCC can be observed for a SOI from $-280\,°CA_{aTDCf}$ to $-310\,°CA_{aTDCf}$. The NO_x concentration is relatively high for these OPs and the combustion duration is reduced, especially when compared to the standard spark plug. Contrary to earlier and later SOIs, the mixture in the pre-chamber is rather rich during these four OPs [128].

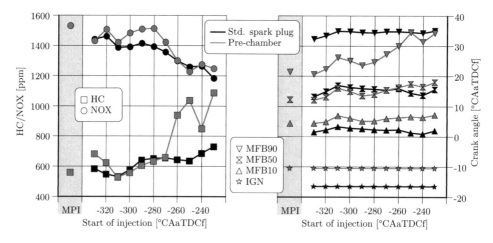

Figure 4.57: Emissions and combustion parameters of standard spark plug and pre-chamber during SOI sweep compared to manifold port injection (MPI) (8 bar IMEP, λ=1, 2000 rpm)

The OP with port fuel injection serves as a reference for a homogeneous mixture. Combustion and emission values are on a low level. However, as the comparison is not done on the knock limit of the engine, the beneficial effect of a gas temperature reduction of the direct injection, due to the fuel evaporation inside the combustion chamber, is not taken into account.

If the load of the engine is reduced to 5 bar IMEP, the combustion with the standard spark plug is still less dependent on the injection timing, while the pre-chamber shows a degradation for later injection timings than $-280\,°CA_{aTDCf}$ in Figure 4.58. Main effect is hereby again the increase in HC and the reduction of NO_x as a result of a less complete energy transformation during a slower combustion. It is remarkable that even though less fuel is injected – than during the 8 bar IMEP OPs – the HC emission is on the same 600 ppm level. An explanation therefore is that the origin of the hydrocarbons is not only the fuel, but also the oil consumption of the engine. The relative comparison between both ignition systems is therefore still valid. The NO_x emissions are dependent on the duration of the combustion and its crank angle position during the cycle, with a stronger weighting on the latter. The pre-chamber has a slight earlier MFB50 and a faster combustion for SOIs from $-340\,°CA_{aTDCf}$ to $-290\,°CA_{aTDCf}$, thus higher NO_x rates can be observed. At $-280\,°CA_{aTDCf}$ the combustion with either a standard spark plug or a pre-chamber is similar with equivalent NO_x emission.

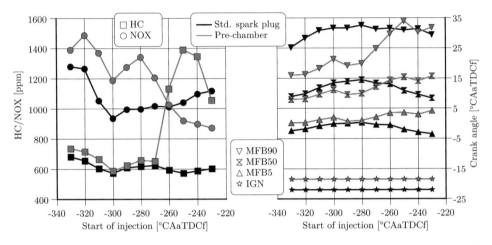

Figure 4.58: Emissions of std. spark plug and pre-camber during SOI sweep together with combustion parameters (4 bar IMEP, λ=1, 2000 rpm)

4.13.2 Measurement of hydrocarbon and nitrogen oxide emissions with different fuels

Except if hydrogen is used as fuel, the emission of CO_2 is inevitable for an ICE. However the origin of the carbon (C) atoms is relevant. Today's fuel ports mainly C of a fossil origin, which was bound over years in organisms and transformed due to pressure and time into oil. The release of this carbon is the critical point of todays ICEs, but also the main driver of engine efficiency development. The creation of fuels that use alternative C sources is addressed to the engine and fuel industry [117]. One candidate of those synthetic fuels or advanced biofuels is the bioliq fuel, which was developed at the KIT [30].

The characteristics – e.g. the higher evaporation temperature and the much higher aromatics content of this fuel synthsis experiment – of the bioliq fuel differs significantly from standard RON95 E5 [86]. To match the DIN EN 228 standard for automotive fuels, 10 Vol% bioliq are blended with 90 Vol% of RON95 E5 fuel. This bioliq/10 blend is then compared to RON95 E5 in the IFKM research engine with a pre-chamber combustion. Figure 4.59 shows the combustion results for both ignition systems and fuels at 8 bar

IMEP and 2000 rpm at λ=1. When a standard spark plug is used, the combustion duration is slightly shorter with the bioliq blend. For the same ignition timing, the MFB50 is at similar $°CA_{aTDCf}$. HC emission is increased during every OP, indicating a less complete combustion with bioliq. However, the NO_x in the exhaust is higher due to the faster combustion and higher octane number.

Contrary to the combustion with the standard spark plug, the bioliq blend has a negative impact on pre-chamber combustion. The combustion duration is longer and the MFB50 is approximately 1 °CA later in the cycle. Like for the standard spark plug, the HC emission is increased. Even though the combustion is slower and later, the NO_x emissions are higher. The most probable reason for this is a higher combustion temperature, due to the lack of the cooling effect during the evaporation of the fuel in the cylinder with the bioliq blend [86].

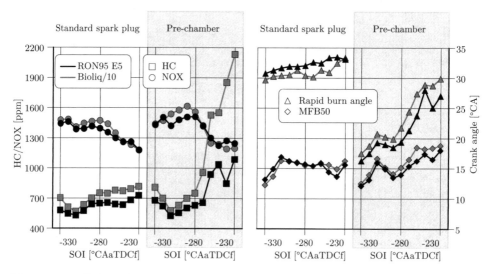

Figure 4.59: Hydrocarbon and Nitrogen oxide emission together with combustion results (MFB50 and rapid burn angle) of RON95 E5 and Bioliq for a standard spark plug and a pre-chamber (8 bar IMEP, λ=1, 2000 rpm)

The IMEP reduction to 5 bar supports the same conclusions as for the standard spark plug at higher load. This means that the bioliq is beneficial for its combustion efficiency. Contrary to the 8 bar IMEP OPs, where the ignition timing is held constant for each ignition system, the spark timing is adjusted to the same MFB50 during the 4 bar IMEP comparison (Figure 4.60). The ignition timing is therefore increased by 1 °CA to $-18.5 °CA_{aTDCf}$ for RON95 E5 during the SOI sweep for the pre-chamber spark plug.

This adjustment matches the observation of the standard spark plug result, where the bioliq blend shows the tendency to an earlier combustion. However, it is contrary to the pre-chamber results at 8bar, where the bioliq blend has a slower combustion.

Once the combustion is adjusted to the same CA position, the NO_x exhaust content of the standard spark plugs is higher than with the pre-chamber. The latter shows a very unstable combustion for SOIs later than $-270\,°\text{CA}_{\text{aTDCf}}$, with high HC emissions, caused by an insufficient fuel supply into the pre-chamber.

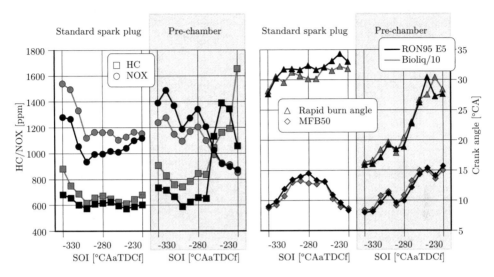

Figure 4.60: Hydrocarbon and Nitrogen oxide emission together with combustion results (MFB50 and rapid burn angle) of RON95 E5 and Bioliq for a standard spark plug and a pre-chamber (4 bar IMEP, λ=1, 2000 rpm)

The experiments confirm the strong impact of the mixture preparation on the pre-chamber ignition. Besides the start of injection – with its direct impact on the fuel support of the pre-chamber – the fuel characteristic is important when the bioliq blend is slowing down the combustion at the higher load OPs with more injected fuel. A reduction of air and fuel demonstrates to be beneficial for the filling of the pre-chamber with the latter. The cause therefore can be found in different mixture formations that are caused by the fuels and their evaporation behavior. An explanation for the divergent behaviors of 8 bar and 5 bar IMEP could be a more concentrated fuel in the center of the MCC for the lower load bioliq OPs. As the bioliq of this synthesis test campaign evaporated less easily, the mixture stays closer to the central mounted injector, resulting in a richer mixture in the pre-chamber.

5 Summary and conclusion

The objective during the research activity was the increase of engine efficiency by the use of a pre-chamber ignition. Main aspects – which influence the pre-chamber behavior – were defined as: mixture composition, temperature, pressure and air motion inside the pre-chamber in interaction with the pre-chamber spark plug design. Different techniques were developed and adapted to increase the knowledge about the spark formation and the pre-chamber combustion process. Besides the mechanical development the necessary change of the engine control mechanisms was investigated.

Various pre-chambers were tested in Viry-Châtillon and Karlsruhe to evaluate the impact of the orifice design on the MCC combustion but also on the internal gas motion inside the spark plug. Especially the combination of hole offset and orifice surface area are key for a proper working system in regards of engine performance. The research revealed that the inflowing air movement is crucial for the pre-chamber performance. To quantify its impact on the spark formation, a spark deflection test chamber was developed at the IFKM during the research. The test chamber helps to evaluate and compare different orifice and spark plug designs in regards of the provoked air movement between the electrodes. A small internal diameter of the pre-chamber and an increased hole offset showed hereby the highest spark deflection. The testing device can also evaluate the impact of wear of the pre-chamber cap or electrodes and clogging of holes on the spark formation.

The pressure measurement inside the pre-chamber proved to be a valuable tool for the understanding of the gas exchange between both volumes. A measurement pre-chamber was designed for the M12 spark plug thread of the IFKM research engine, while a modified cylinder head was needed at the RSR facility. The knowledge about the pressure evolution in the pre-chamber and the MCC in parallel helps to find the correct relation between pre-chamber volume and air-fuel ratio for an optimal OP. It supports also to determine and define the boundaries for the mechanical development of the pre-chamber components and to identify the origin of engine malfunctions.

To know the temperature at relevant points, such as the pre-chamber cap and the ground electrode, reveals hot spots and possible origins for surface ignition. Supported by thermal simulation and the pressure acquisition, the center electrode was determined as the most probable cause for surface ignition. The impact of fuel mass and air-fuel ratio as the main driver for the spark plug temperature were discovered.

Based on the results of temperature and pressure measurement, different materials for the pre-chamber cap were chosen and compared in the engine. Copper alloy proves to be vulnerable for wear during the necessary lifetime. Nickel alloys show a more constant performance over time but at the cost of a higher cap temperature. Other promising design subjects like a pre-chamber spark plug with a non-resistor insulator assembly, a converging-diverging nozzle design in the pre-chamber cap and the thermal insulation of the pre-chamber were tested but did not show the expected benefits.

To create further knowledge, two optical analyzing methods were used with the IFKM engine to analyze mixture preparation inside the pre-chamber. An endoscopic high speed acquisition served to determine the spark deflection and to connect the pressure with jet ejection into the MCC. As second technique an IR sensor measured the mixture composition inside the pre-chamber. Due to absorption of HC and CO_2 molecules, the identification of unburned and burned fuel inside the pre-chamber was possible. The unique measurement pre-chamber spark plug was designed in the research process of the present work. Insight about the local air-fuel ratio helps to evaluate the effect of a richer pre-chamber on the MCC combustion, for example during a SOI variation. A faster combustion can therefore be connected to a richer pre-chamber. A surprising result of the experiments is the gas composition inside the pre-chamber after combustion. During the expansion stroke unburned fuel is detected inside the pre-chamber causing the assumption that its volume is filled with fresh mixture during the early pressure increase in the MCC. The differential pressure between pre-chamber and MCC shows a refilling during the pressure rise in the latter. One hypothesis for this observation is a refilling with fresh mixture during this period of time. Another potential explanation for the origin of the measured HCs is quenching of the combustion on the walls inside the pre-chamber. Further detailed investigations and specialized measurement setups are necessary to determine the cause for the unburden fuel in the pre-chamber after the main chamber combustion.

The evolution of the carbon dioxide inside the pre-chamber was a finding in contrary to theses in known publications, as its concentration decreases after the combustion and does not stay on a constant level as expected. An observation that is contrary to common simulation results, where the pre-chamber is filled entirely with burned gas at the beginning of the compression stroke. The CO_2 measurement inside the pre-chamber is only possible due to the integration of the sensor in a functioning pre-chamber in a fired engine.

Additional measurement of HC and NO_x emissions in the exhaust gas for the pre-chamber – in comparison to a standard spark plug – confirms the great dependence of the pre-chamber on the SOI of the direct injector. However, it was also evaluated that the sensor delivers coherent HC information with fuel injection in a motored engine, to evaluate the energy supply into the pre-chamber. Valuable information for its use on motored high load points that are not supported in a fired engine by the fragile measurement technology.

Pre-chamber combustion reduces the flame development angle and the combustion duration, resulting in a faster pressure rise in the combustion chamber. The increased amplitude of the high pass filtered cylinder pressure signal might be wrongly connected to a knock event. Thus control mechanisms of the engine need to be adapted to the increased noise on the cylinder pressure signal. Additionally it was found that engines become more dependent on the humidity of the combustion air when a pre-chamber is used. The research showed that an increase of the pre-chamber volume helps to compensate the loss of combustion stability that occurs with a higher water content in the combustion air.

A guideline for the development of a pre-chamber ignition that is based on the findings of the present work is displayed in Figure 5.1. The development is hereby divided into five phases:

The first phase focuses on the concept choice between a pre-chamber insert and a pre-chamber spark plug. Main input therefore are cylinder head geometry and the need of an active fuel supply. As pre-chamber spark plugs are easier to change, they are beneficial if multiple orifice and volumes should be tested. Thereafter, a pre-chamber is designed that will serve for the first engine tests. Simulation, literature and experience are hereby key. The objective of the third phase is a reliable pre-chamber design. Especially full load operation and spark formation is critical. Once the design is validated the optimization of the pre-chamber follows in the fourth phase, where simulation and experiments are intensified until lifetime, OPs and performance targets are reached. The last development step is the optimization of the engine with and around the new combustion system.

The different loops during the development can be repeated and adapted with the evolving engine. The experience gained from motorsport in this project showed so far no end and continuously benefits for the engine efficiency by the optimization of the pre-chamber ignition.

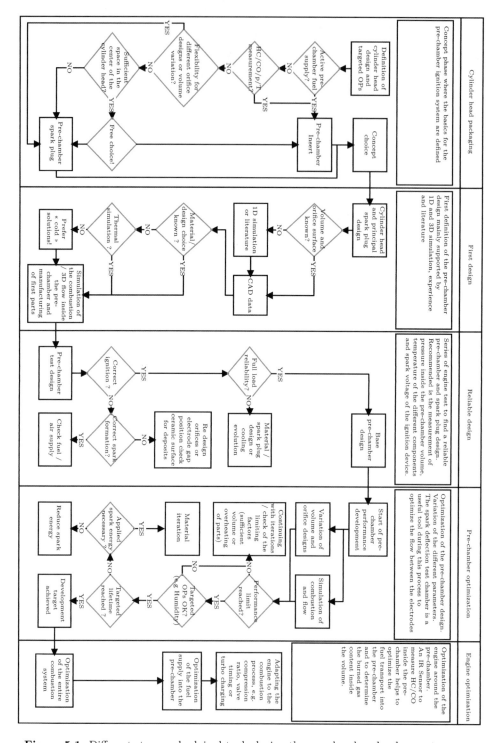

Figure 5.1: Different steps and advised tools during the pre-chamber development process

Nomenclature

Physical Quantities

Symbol	Unit	Description
A	-	Fitting Parameter (U_B calculation)
a	W/mK	Thermal conductivity
α	-	First Townsend coefficient
$\alpha\, p_{Cyl}\, max$	°CAaTDCf	Crank angle position of maximal cylinder pressure
B	-	Fitting Parameter (U_B calculation)
$BSFC$	g/kWh	Break specific fuel consumption
β	°	Angle between pre-chamber holes
C_{Cable}	F	Ignition cable capacity
C_{Coil}	F	Ignition coil capacity
C_{SP1}	F	First spark plug capacity
C_{SP2}	F	Second spark plug capacity
c_p	J/K	Heat capacity at constant pressure
c_v	J/K	Heat capacity at constant volume
CA	°	Crank angle
CA_{aTDCf}	°	Crank angle after top dead center firing
CA_{aSI}	°	Crank angle after spark ignition
CO_2	%	Carbon dioxide concentration
COV_{IMEP}	%	Coefficient of variation of IMEP
d	mm	Pre-chamber hole diameter
$\Delta pressure$	bar	Pressure difference between pre-chamber and main combustion chamber
E_A	mJ	Arc energy
E_B	mJ	Breakdown energy
E_G	mJ	Glow energy
E_{Spark}	mJ	Spark energy
ε	-	Compression ratio
η_f	%	Fuel efficiency
η_i	%	Indicated efficiency
η_m	%	Mechanical efficiency
$\eta_{th,C}$	%	Thermal efficiency Carnot cycle
$\eta_{th,m}$	%	Thermal efficiency mixed cycle
$\eta_{th,v}$	%	Thermal efficiency constant volume cycle
Fda	°CA	Flame development angle
γ	-	Third Townsend coefficient
HC	ppm	Hydro carbon

Symbol	Unit	Description
$I_{iA.0}$	A	Breakdown current
$I_{iA.1}$	A	Unsteady arc current
$I_{iA.2}$	A	Arc current
I_G	A	Glow current
I_{Spark}	A	Spark current
Ign	°CA	Ignition timing
$IMEP$	bar	Indicated mean effective pressure
Kp	bar	Knock pressure
κ	-	Heat capacity ratio
L_{Prim}	H	Primary coil inductance
L_{Sec}	H	Secondary coil inductance
l_{spark}	mm	Spark elongation
λ	-	Air-fuel equivalence ratio
$MFB10$	°CA	10% Mass fraction burned
$MFB50$	°CA	50% Mass fraction burned
$MFB90$	°CA	90% Mass fraction burned
$\dot{m}_{air}$	kg/h	Air mass flow
m_f	kg	Fuel mass
$\dot{m}_{fuel}$	kg/h	Fuel mass flow
m_{PC}	g	Gas mas in pre-chamber
n	-	Number of holes
NO_x	ppm	Nitrogen oxide
o	mm	Offset of pre-chamber hole
p	bar	Pressure
p_{Cyl}	bar	Cylinder pressure
$p_{Cyl}max$	bar	Maximal cylinder pressure
p_{PC}	bar	Pressure in pre-chamber
p_{tch}	bar	Pressure in spark deflection test chamber
Q_{HV}	kJ/kg	Heating value
q_{rem}	J	Removed heat
q_{suppl}	J	Supplied heat
R	J/(K mol)	Ideal gas constant
R_{Coil}	Ohm	Spark coil resistance
R_{Prim}	Ohm	Primary coil resistance
R_{SP}	Ohm	Spark plug resistance
ρ_u	kg/m^3	Density of unburned gas
ρ_f	kg/m^3	Fuel density
s	J/kg	Entropy
S_t	m^2	Surface of the turbulent flame front
S_{fl}	m^2	Surface of the laminar flame front
sH	g/kg	Specific humidity
SOI	°CAaTDCf	Start of injection
σ	-	Standard derivation
T	°C	Temperature
T_{min}	K	Minimal temperature
T_{max}	K	Maximal temperature

Symbol	Unit	Description
T_{Man}	°C	Manifold air temperature
T_{PC}	°C	Gas temperature in pre-chamber
T_PC	°C	Pre-chamber temperature
T_SP	°C	Spark plug temperature
T_W	°C	Water temperature
τ_{ch}	s	Chemical reaction time
U_A	kV	Arc voltage
U_{iA}	kV	Unsteady arc voltage
U_B	kV	Breakdown voltage
U_G	kV	Glow voltage
U_{Spark}	kV	Spark voltage
V	mm³	Volume
V_{PC}	mm³	Pre-chamber volume
v'	m/s	Turbulent fluctuation flame speed
v_{fl}	m/s	Laminar flame speed
v_{spark}	m/s	Spark deflection velocity
v_t	m/s	Turbulent flame speed
W_e	J	Effective work
W_i	J	Indicated work
x_b	-	Burned gas fraction

Abbreviations and indices

Abbreviation	Description
BMEP	Break mean effective pressure
BDC	Bottom dead center
BSFC	Break specific fuel consumption
CAD	Computer aided design
CAI	Controlled auto ignition
CD	Converging-diverging
CH	Methylidyne
CO	Carbon monoxide
COV	Coefficient of variation
CU	Copper
DDI	Direct diesel injection
DFI	Direct fuel injection
DIN	Deutsche Industrie Norm
ED	Electrode distance
F1	Formula 1
FIA	Fédération Internationale de l'Automobile
HC	Hydrocarbon
ICE	Internal combustion engine

Abbreviation	**Description**
ICOS	Internal combustion optical sensor
IFKM	Institute of combustion engines
IR	Infrared
IVC	Intake valve closing
KIT	Karlsruhe Institute of Technology
Lambda	Air-fuel equivalence ratio
MCC	Main combustion chamber
MPI	Manifold port injection
MSP	Multiple spark point
NiCu	Nickel copper alloy
NiCr	Nickel chrome alloy
OH	Hydroxide
OP	Operation point
RSR	Renault Sport Racing
SI	Spark ignition
SGSP	Single gap spark plug
SP	Spark plug
TKE	Turbulent kinetic energy
TDC	Top dead center
TJI	Turbulent jet ignition

Bibliography

[1] ALBRECHT H. ; MALY R. ; SAGGAU B. ; WAGNER E.: Neue Erkenntnisse ueber elektrische Zuendfunken und ihre Eignung zur Entflammung brennbarer Gemische Teil1. In: Automobil-Industrie (1977), No. 04-1977, pp. 45–50

[2] ALONSO, L. ; OREGUI, I. ; WEINROTTER, M. ; IRURETAGOIENA, I. : Pre-chamber spark plug development for highest efficiencies at Dresser-Rand's Guascor gas engines. In: Siebenpfeiffer W. (eds) Heavy-Duty-, On- und Off-Highway-Motoren 2014 (2019), pp. 87–103. – DOI 10.1007/978–3–658–23789–9_6

[3] ASSANIS, D. N. ; MATHUR, T. : The Effect of Thin Ceramic Coatings on Spark-Ignition Engine Performance. In: SAE Technical Paper 900903 (1990). – DOI 10.4271/900903

[4] ATTARD, W. P.: Combustion System For Spark Turbulentjet Ignition Pre-chamber Ignition Engines. In: United States Patent Application Publication (2012), No. US 2012/0103302A1

[5] ATTARD, W. P. ; FRASER, N. ; PARSONS, P. ; TOULSON, E. : A Turbulent Jet Ignition Pre-Chamber Combustion System for Large Fuel Economy Improvements in a Modern Vehicle Powertrain. In: SAE International Journal of Engines 3 (2010), No. 2, pp. 20–37. – DOI 10.4271/2010–01–1457. – ISSN 1946–3944

[6] ATTARD, W. P. ; KOHN, J. ; PARSONS, P. : Ignition Energy Development for a Spark Initiated Combustion System Capable of High Load, High Efficiency and Near Zero NOx Emissions. In: SAE International Journal of Engines 3 (2010), No. 2, pp. 481–496. – DOI 10.4271/2010–32–0088. – ISSN 1946–3944

[7] ATTARD, W. P. ; PARSONS, P. : A Normally Aspirated Spark Initiated Combustion System Capable of High Load, High Efficiency and Near Zero NOx Emissions in a Modern Vehicle Powertrain. In: SAE International Journal of Engines 3 (2010), No. 2, pp. 269–287. – DOI 10.4271/2010–01–2196. – ISSN 1946–3944

[8] ATTARD, W. P. ; TOULSON, E. ; HUISJEN, A. ; CHEN, X. ; ZHU, G. ; SCHOCK, H. : Spark Ignition and Pre-Chamber Turbulent Jet Ignition Combustion Visualization. In: SAE International (2012), No. 2012-01-0823. – DOI 10.4271/2012–01–0823

[9] AVL LIST GMBH: Engine Indicating User Handbook. Graz : AVL List GmbH, 2002

[10] BABRAUSKAS, V. : Ignition of Gases, Vapors, and Liquids by Hot Surfaces. In: International Symposium on Fire Investigation Science and Technology (2008), pp. 4–14

[11] BALLAL, D. R. ; LEFEBVRE, A. H.: The influence of flow parameters on minimum ignition energy and quenching distance. In: Symposium (International) on Combustion 15 (1975), No. 1, pp. 1473–1481. – DOI 10.1016/S0082–0784(75)80405–X. – ISSN 00820784

[12] BARDIS, K. ; XU, G. ; KYRTATOS, P. ; WRIGHT, Y. M. ; BOULOUCHOS, K. : A Zero Dimensional Turbulence and Heat Transfer Phenomenological Model for Pre-Chamber Gas Engines. In: SAE Technical Paper 2018-01-1453 (2018). – DOI 10.4271/2018–01–1453

[13] BAUMGARTNER, L. S. ; KARMANN, S. ; BACKES, F. ; STADLER, A. ; WACHT-MEISTER, G. : Experimental Investigation of Orifice Design Effects on a Methane Fuelled Prechamber Gas Engine for Automotive Applications. In: SAE Technical Paper 2017-24-0096 (2017). – DOI 10.4271/2017–24–0096

[14] BEDOGNI, F. ; MAGISTRALI, S. ; MAZZONI, D. ; MUSUS, E. ; PIVETTI, G. ; ZOLESI, P. : Gasoline Internal Combustion Engine, With A Combustion Pre-chamber And Two Spark Plugs. EP 3 453 856 B1 (2019)

[15] BENAJES, J. ; NOVELLA, R. ; GOMEZ-SORIANO, J. ; MARTINEZ-HERNANDIZ, P. J. ; LIBERT, C. ; DABIRI, M. : Performance of the passive pre-chamber ignition concept in a spark-ignition engine for passenger car applications. In: Proceedings of the SIA Powertrain Electronics (12–13 June 2019), pp. 1–10

[16] BERG, T. ; THIELE, O. ; SEEFELDT, S. ; VANHAELST, R. : Bestimmung der innermotorischen Gemischbildung durch optisches Indizieren. In: MTZ - Motortechnische Zeitschrift 74 (2013), No. 6, pp. 472–477. – DOI 10.1007/s35146–013–0140–4. – ISSN 0024–8525

[17] BISWAS, S. ; QIAO, L. : Prechamber Hot Jet Ignition of Ultra-Lean H 2 /Air Mixtures: Effect of Supersonic Jets and Combustion Instability. In: SAE International Journal of Engines 9 (2016), No. 3. – DOI 10.4271/2016–01–0795. – ISSN 1946–3944

[18] BISWAS, S. ; QIAO, L. : A Numerical Investigation of Ignition of Ultra-Lean Premixed H 2 /Air Mixtures by Pre-Chamber Supersonic Hot Jet. In: SAE International Journal of Engines 10 (2017), No. 5. – DOI 10.4271/2017–01–9284. – ISSN 1946–3944

[19] BLANKMEISTER, M. ; ALP, M. ; SHIMIZU, E. : Passive Pre-Chamber Spark Plug for Future Gasoline Combustion Systems with Direct Injection. In: Günther, M., Sens M.(eds) Ignition Systems for Gasoline Engines : 4th International Conference. Proceedings. expert verlag GmbH, Tübingen. (2018), pp. 149–174. – DOI 10.5445/IR/1000088324

[20] BRYZIK, W. ; KAMO, R. : TACOM/Cummins Adiabatic Engine Program. In: SAE Technical Paper 830314 (1983). – DOI 10.4271/830314

[21] BUNCE, M. ; BLAXILL, H. : Sub-200 g/kWh BSFC on a Light Duty Gasoline Engine. In: SAE Technical Paper 2016-01-0709 (2016). – DOI 10.4271/2016–01–0709

[22] BUNCE, M. ; BLAXILL, H. ; KULATILAKA, W. ; JIANG, N. : The Effects of Turbulent Jet Characteristics on Engine Performance Using a Pre-Chamber Combustor. In: SAE Technical Paper 2014-01-119 (2014). – DOI 10.4271/2014–01–1195

[23] BURGETT, R. R. ; MASSOLL, R. E. ; VAN UUM, D. R.: Relationship Between Spark Plugs and Engine-Radiated Electromagnetic Interference. In: SAE Technical Paper 740111 (1974). – DOI 10.4271/740111

[24] CARNOT, S. : Réflexions sur la puissance motrice du feu et sur les machines propres à développer cette puissance. In: Annales scientifiques de l'École normale supérieure 1 (1872), pp. 393–457. – DOI 10.24033/asens.88. – ISSN 0012–9593

[25] CHIODI, M. ; KAECHELE, A. ; BARGENDE, M. ; WICHELHAUS, D. ; POETSCH, C. : Development of an Innovative Combustion Process: Spark-Assisted Compression Ignition. In: SAE International Journal of Engines 10 (2017), No. 5. – DOI 10.4271/2017–24–0147. – ISSN 1946–3944

[26] CONTI, M. ; KLUZER, D. : Maserati MC20: the Brand's new super sports car. https://www.media.maserati.com/assets/documents/original/16863-PRESSRELEASEMC20DEFENG.docx. Version: 9.9.2020

[27] CORRIGAN, D. J. ; DI SACCO, M. ; MEDDA, M. ; PALTRINIERI, S. ; ROSSI, V. : High - Performance Internal Combustion Engine with Improved Handling of Emission and Method of Controlling such Engine. In: United States Patent Application Publication US 2019 / 0323415 A1 (2019), No. US 2019 / 0323415 A1

[28] COWARD, H. F. ; GUEST, P. G.: Ignition of Natural Gas-air Mixtures by Heated Metal Bars. In: Journal of the American Chemical Society 49 (1927), No. 10, pp. 2479–2486. – DOI 10.1021/ja01409a017. – ISSN 0002–7863

[29] COWART, J. S. ; HAMILTON, L. J.: Fuel Accounting Analysis during Cranking and Startup using Simultaneous In-Cylinder and Exhaust Fast FID and NDIR Detectors. In: SAE International Journal of Engines 1 (2009), No. 1, pp. 820–830. – DOI 10.4271/2008–01–1309. – ISSN 1946–3944

[30] DAHMEN, N. ; ABELN, J. ; EBERHARD, M. ; KOLB, T. ; LEIBOLD, H. ; SAUER, J. ; STAPF, D. ; ZIMMERLIN, B. : The bioliq process for producing synthetic transportation fuels. In: Wiley Interdisciplinary Reviews: Energy and Environment 6 (2017), No. 3, pp. e236. – DOI 10.1002/wene.236. – ISSN 20418396

[31] DATE, T. ; NOMURA, T. : Research and Development of the Carburetor for the CVCC Engine. In: SAE Technical Paper 800507 (1980). – DOI 10.4271/800507

[32] DAVIS, S. ; KELLY, S. ; SOMANDEPALLI, V. : Hot Surface Ignition of Performance Fuels. In: Fire Technology 46 (2010), No. 2, pp. 363–374. – DOI 10.1007/s10694–009–0082–z. – ISSN 0015–2684

[33] DEUTSCHER MOTOR SPORT BUND E.V.: DTM-Technisches-Reglement-2020. https://www.motorsport-total.com/styles/mst/media/f1-reglement/DTM-Technisches-Reglement-2020.pdf. Version: 2020 Access date: 30.01.2023

[34] DIETSCHE, K.-H. :   Automotive handbook.   9th edition, revised and extended. Karlsruhe : Robert Bosch GmbH, September 2014. – ISBN 9780768081527

[35] DISCH, C. ; HUEGEL, P. ; BUSCH, S. ; KUBACH, H. ; SPICHER, U. ; PFEIL, J. ; DIRUMDAM, B. ; WALDENMAIER, U. : High-Speed Flame Chemiluminescence Investigations of Prechamber Jets in a Lean Mixture Large-Bore Natural Gas Engine. In: 27th CIMAC World Congress Paper (2013), No. 79, pp. 1–16

[36] ENDO, H. ; TAKI, M. ; OIKAWA, Y. :   Internal Combustion engine. In: United States Patent Application Publication US 2019/0203636 A1 (2019)

[37] EUROPEAN ENVIRONMENT AGENCY ; EUROPEAN ENVIRONMENT AGENCY (eds.): This figure presents the fuel efficiency and fuel consumption trends for private cars in the EU-28 in the period 1990 to 2015. The variables included are number of cars, average CO2 emissions of cars, average fuel consumption of cars, GDP, total distance travelled by cars and total energy consumptions of cars. https://www.eea.europa. eu/ds_resolveuid/e5ff6a1e366246368efbb15a4545599b.   Version: 16.12.2019, Access date: 16.04.2020

[38] FÉDÉRATION INTERNATIONALE DE L'AUTOMOBILE: 2013 Formula One Technical Regulations.   https://www.fia.com/sites/default/files/regulation/file/ 2013%20F1TECHNICAL%20REGULATIONS%20-%20PUBLISHED%20ON%2004.07.2013. pdf%20ON%2004.07.2013.pdf. Version: 2013

[39] FÉDÉRATION INTERNATIONALE DE L'AUTOMOBILE: 2014 Technical Regulations for LMP1 Prototype.   https://www.fia.com/sites/default/files/ regulation/file/TECHNICAL%20REGULATIONS%20LMP1%20%282014%29-V08-27. 09.2013-FINAL.pdf. Version: 2014

[40] FÉDÉRATION INTERNATIONALE DE L'AUTOMOBILE: 2018 Formula One Technical Regulations. https://www.fia.com/sites/default/files/1-2018_technical_ regulations_2017-12-19_0.pdf. Version: 2017

[41] FISCHER, M. ; GÜNTHER, M. ; RÖPKE, K. ; LINDEMANN, M. ; PLACEZK, R. : Klopferkennung im Ottomotor: Neue Tools und Methoden in der Serienentwicklung. In: MTZ Motortechnische Zeitschrift (2003), No. 64, pp. 186–195

[42] GOLLOCH, R. : Downsizing bei Verbrennungsmotoren: Senkung des Kraftstoffverbrauchs und Steigerung des Wirkungsgrads. 1. Aufl. Berlin : Springer, 2005 (VDI). – ISBN 3540238832

[43] GÖRGEN, M. ; BALAZS, A. ; BÖHMER, M. ; NIJS, M. ; LEHN, H. ; SCHARF, J. ; THEWES, M. ; MÜLLER, A. ; ALT, N. ; CLASSEN, J. ; STERLEPPER, S. : All lambda 1 gasoline powertrains. In: Liebl J., Beidl C., Maus W. (eds) Internationaler Motorenkongress 2018. Proceedings. Springer Vieweg, Wiesbaden. (2018), pp. 93– 111. – DOI 10.1007/978-3-658-21015-1_7

[44] GROSCH, A. ; WACKERBARTH, H. ; THIELE, O. ; BERG, T. ; BECKMANN, L. : Infrared spectroscopic concentration measurements of carbon dioxide and gaseous water in harsh environments with a fiber optical sensor by using the HITEMP database. In: Journal of Quantitative Spectroscopy and Radiative Transfer 133 (2014), pp. 106–116. – DOI 10.1016/j.jqsrt.2013.07.021. – ISSN 00224073

[45] GROSCH, A. ; BEUSHAUSEN, V. ; THIELE, O. ; GRZESZIK, R. : Crank Angle Resolved Determination of Fuel Concentration and Air/Fuel Ratio in a SI-Internal Combustion Engine Using a Modified Optical Spark Plug. In: SAE Technical Paper 2007-01-0644 (2007). – DOI 10.4271/2007–01–0644

[46] GÜNTHER, M. (eds.) ; SENS, M. (eds.): Ignition Systems for Gasoline Engines: 3rd International Conference, November 3-4, 2016, Berlin, Germany. Cham : Springer International Publishing, 2017. – ISBN 978–3–319–45503–7

[47] GÜNTHER, M. (eds.) ; SENS, M. (eds.): Ignition Systems for Gasoline Engines: 4th International Conference, December 6-7, 2018, Berlin, Germany. Tübingen : expert verlag GmbH, 2018. – ISBN 978–3–8169–3449–3

[48] GUSSAK, L. A. ; KARPOV, V. P. ; TIKHONOV, Y. V.: The Application of Lag-Process in Prechamber Engines. In: SAE Technical Paper 790692 (1979). – DOI 10.4271/790692

[49] GUSSAK, L. A. ; TURKISH, M. C. ; SIEGLA, D. C.: High Chemical Activity of Incomplete Combustion Products and a Method of Prechamber Torch Ignition for Avalanche Activation of Combustion in Internal Combustion Engines. In: SAE Technical Paper 750890 (1975). – DOI 10.4271/750890

[50] HABERMANN, K. ; MORCINKOWSKI, B. ; MÜLLER, C. ; SCHERNUS, C. ; UHLMANN, T. : Development of a pre-chamber for ultra-lean operation of gasoline engines. In: Proceedings of 8th Transport Research Arena TRA 2020, April 27-30, 2020, Helsinki, Finland, Apr 2020, Helsinki, Finland. (2020)

[51] HALL, M. J. ; KOENIG, M. : A fiber-optic probe to measure precombustion in-cylinder fuel-air ratio fluctuations in production engines. In: Symposium (International) on Combustion 26 (1996), No. 2, pp. 2613–2618. – DOI 10.1016/S0082–0784(96)80095–6. – ISSN 00820784

[52] HALL, M. J. ; ZUZEK, P. ; ANDERSON, R. W.: Fiber Optic Sensor for Crank Angle Resolved Measurements of Burned Gas Residual Fraction in the Cylinder of an SI Engine. In: SAE Technical Paper 2001-01-1921 (2001). – DOI 10.4271/2001–01–1921

[53] HEINZ, C. E.: Untersuchung eines Vorkammerzündkonzepts für Großgasmotoren in einer Hochdruckzelle mit repetierender Verbrennung, Technische Universität München, Dissertation, 2011

[54] HELMETSBERGER, P. : Experimentelle Gemischbildungsuntersuchungen an einem Ottomotor mit vollvariablem Ventiltrieb, Direkteinspritzung und Aufladung., Technische Universität Graz, Dissertation, 2010

[55] HEYWOOD, J. B.: Internal Combustion Engine Fundamentals. United States of America : McGraw-Hill Education and McGraw-Hill, 1988 (McGraw-Hill series in mechanical engineering). – ISBN 0–07–028637–X

[56] HÜCHTEBROCK, B. ; GEIGER, J. ; DHONGDE, A. ; SANKHLA, H. : Development of a Natural Gas Combustion System for High Specific Power. In: MTZ Worldwide, Wiesbaden, Germany 76 (2015), No. 10, pp. 30–35. – DOI 10.1007/s38313–015–0043–5

[57] JANAS, P. ; NIESSNER, W. : Towards a Thermally Robust Automotive Pre-Chamber Spark Plug for Turbocharged Direct Injection Gasoline Engines. In: Günther, M., Sens M.(eds) Ignition Systems for Gasoline Engines : 4th International Conference. Proceedings. expert verlag GmbH, Tübingen. (2018), pp. 122–148

[58] JAPAN AUTOMOBILE FEDERATION: 2016 JAF Grand Touring Car 500 (JAF-GT500) Technical Regulations. http://www.jaf.or.jp/msports/rules/image/2016JAF-GT500_en.pdf. Version: 2016

[59] JAROSINSKI, J. ; LAPUCHA, R. ; MAZURKIEWICZ, J. ; WOJCICKI, S. : Combustion System of a Lean-Burn Piston Engine with Catalytic Prechamber. In: SAE Technical Paper 2001-01-1186 (2001). – DOI 10.4271/2001–01–1186

[60] JEFFRIES, J. B. ; PORTER, J. M. ; PYUN, S. H. ; HANSON, R. K. ; SHOLES, K. R. ; SHOUJI, K. ; CHAYA, T. :   An In-cylinder Laser Absorption Sensor for Crank-angle-resolved Measurements of Gasoline Concentration and Temperature. In: SAE International 2010-01-2251 (2010). – DOI 10.4271/2010–01–2251

[61] KALLMEYER, F. : Methoden zur Untersuchung der lokalen Gemischzusammensetzung im DI-Ottomotor, Universität Duisburg-Essen, Dissertation, 2009

[62] KAWABATA, Y. ; MORI, D. : Combustion Diagnostics & Improvement of a Prechamber Lean-Burn Natural Gas Engine. In: SAE Technical Paper 2004-01-0979 (2004). – DOI 10.4271/2004–01–0979

[63] KAWAGUCHI, A. ; WAKISAKA, Y. ; NISHIKAWA, N. ; KOSAKA, H. ; YAMASHITA, H. ; YAMASHITA, C. ; IGUMA, H. ; FUKUI, K. ; TAKADA, N. ; TOMODA, T. : Thermo-swing insulation to reduce heat loss from the combustion chamber wall of a diesel engine. In: International Journal of Engine Research 20 (2019), No. 7, pp. 805–816. – DOI 10.1177/1468087419852013. – ISSN 1468–0874

[64] KETTNER, M. : Experimentelle und numerische Untersuchungen zur Optimierung der Entflammung von mageren Gemischen bei Ottomotoren mit Direkteinspritzung, Universität Karlsruhe, Dissertation, 2006

[65] KETTNER, M. ; ROTHE, M. ; VELJI, A. ; SPICHER, U. ; KUHNERT, D. ; LATSCH, R. : A New Flame Jet Concept to Improve the Inflammation of Lean Burn Mixtures in SI Engines. In: SAE Technical Paper 2005-01-3688 (2005). – DOI 10.4271/2005–01–3688

[66] KIM, J. ; ANDERSON, R. W.: Spark Anemometry of Bulk Gas Velocity at the Plug Gap of a Firing Engine. In: SAE Technical Paper 952459 (2015). – DOI 10.4271/952459

[67] KIM, T. ; SONG, J. ; PARK, S. : Effects of turbulence enhancement on combustion process using a double injection strategy in direct-injection spark-ignition (DISI) gasoline engines. In: International Journal of Heat and Fluid Flow 56 (2015), pp. 124–136. – DOI 10.1016/j.ijheatfluidflow.2015.07.013. – ISSN 0142727X

[68] KIMURA, N. ; KOBAYASHI, H. ; ISHIKAWA, N. : Study of Gasoline Pre-chamber combustion at Lean Operation. In: Günther, M., Sens M.(eds) Ignition Systems for Gasoline Engines : 4th International Conference. Proceedings. expert verlag GmbH, Tübingen. (2018), pp. 275–291. – DOI 10.5445/IR/1000088589

[69] KOENIG, M. ; HALL, M. J.: Measurements of Local In-Cylinder Fuel Concentration Fluctuations in a Firing SI Engine. In: SAE Technical Paper 971644 (1997). – DOI 10.4271/971644

[70] KOENIG, M. H. ; HALL, M. J.: Cycle-Resolved Measurements of Pre-Combustion Fuel Concentration Near the Spark Plug in a Gasoline SI Engine. In: SAE Technical Paper 981053 (1998). – DOI 10.4271/981053

[71] KONSTANTINOFF, L. ; DORNAUER, T. ; STÄRZ, R. ; PILLEI, M. ; MÖLTNER, L. : Untersuchungen zur Ausbildung und Bewegung von Zündfunken in Gasmotoren mit turbulenter Verbrennung. In: 39.Internationales Wiener Motorensymposium (2018), pp. 335–366. – DOI 10.51202/9783186807120

[72] KUCHTA, J. M. ; CATO, R. J.: Hot Gas Ignition Tempratures of Hydrocabon Fuel Vapor-Air Mixtures. In: Bureau of Mines Report of Investigation 6857 (1966)

[73] KUCHTA, J. M.: Investigation of Fire and Explosion Accidents in the Chemical, Mining, and Fuel-Related Industries-A Manual. Washington D.C. : United States. Government Printing Office, 1985

[74] KUHNERT, D. ; LATSCH, R. : Vorkammerzündkerze und Verfahren zur Herstellung derselben. In: European Patent Specification EP 1 476 926 B1 (2006)

[75] LATSCH, R. : The Swirl-Chamber Spark Plug: A Means of Faster, More Uniform Energy Conversion in the Spark-Ignition Engine. In: SAE Technical Paper 840455 (1984). – DOI 10.4271/840455

[76] LAVISION GMBH: ICOS evaluation software: Product-Manual. Göttingen, Germany : LaVision GmbH, 2017

[77] LAVISION GMBH: ICOS M5 Probe: Product Manual. Göttingen, Germany : LaVision GmbH, 2019

[78] LIST, H. ; EICHLSEDER, H. ; KLÜTING, M. ; PIOCK, W. F.: Grundlagen und Technologien des Ottomotors: Der Fahrzeugantrieb. Wien and New York, NY and Heidelberg : Springer, 2008 (Der Fahrzeugantrieb). – ISBN 978–3–211–25774–6

[79] LOVRENICH, R. T. ; HARDIN, J. T.: Electrical to Thermal Conversion in Spark Ignition. In: SAE Technical Paper 670114 (1967). – DOI 10.4271/670114

[80] LUCAS, G. ; TALLU, G. ; WEISSNER, M. : CFD-based Development of an Ignition Chamber for a lean and high efficient CNG Combustion. Valencia, 2018 (THIESEL 2018 Conference on Thermo- and Fluid Dynamic Processes in Direct Injection Engines)

[81] LÜFT, M. ; EICHELDINGER, S. ; NGUYEN, H. D. ; DOHRMANN, S. ; KUPPA, K. : Mitteldruck > 30 bar bei Gasmotoren: Abschlussbericht über das Vorhaben Nr. 1201. In: Informationstagung Motoren, Heft R582 (2018), No. R582, pp. 361–400

[82] MALY, R. ; VOGEL, M. : Initiation and propagation of flame fronts in lean CH4-air mixtures by the three modes of the ignition spark. In: Symposium (International) on Combustion 17 (1979), No. 1, pp. 821–831. – DOI 10.1016/S0082–0784(79)80079–X. – ISSN 00820784

[83] MAUL, G. ; PAESOLD, J. ; SAILER, U. ; KUHNERT, D. : Pre-Chamber Spark Plug. In: United States Patent Application Publication US 8,324,792 B2 (2012)

[84] MERKER, G. P. ; SCHWARZ, C. ; TEICHMANN, R. : Grundlagen Verbrennungsmotoren: Funktionsweise, Simulation, Messtechnik // Funktionsweise, Simulation, Messtechnik ; mit 43 Tabellen. 5., vollständig überarbeitete, aktualisierte und erweiterte Auflage, 2011 // 5., vollständig überarbeitete, aktualisierte und erweiterte Aufl. Wiesbaden : Vieweg+Teubner Verlag / Springer Fachmedien Wiesbaden GmbH Wiesbaden and Vieweg + Teubner, 2012 // 2011 (ATZ/MTZ-Fachbuch). – ISBN 978–3–8348–1393–0

[85] MICHLER, T. ; TOEDTER, O. ; KOCH, T. : Measurement of temporal and spatial resolved rotational temperature in ignition sparks at atmospheric pressure. In: Automotive and Engine Technology 5 (2020), No. 1-2, pp. 57–70. – DOI 10.1007/s41104–020–00059–w. – ISSN 2365–5127

[86] MICHLER, T. ; WIPPERMANN, N. ; TOEDTER, O. ; NIETHAMMER, B. ; OTTO, T. ; ARNOLD, U. ; PITTER, S. ; KOCH, T. ; SAUER, J. : Gasoline from the bioliq® process: Production, characterization and performance. In: Fuel Processing Technology 206 106476 206 (2020). – DOI 10.1016/j.fuproc.2020.106476

[87] MITSUBISHI MOTORS: Mitsubishi Motors Adds World First V6 3.5-liter GDI Engine to Ultra-efficiency GDI Series. https://web.archive.org/web/20091001184522/http://media.mitsubishi-motors.com/pressrelease/e/corporate/detail215.html. Version: 16.04.1997, Access date: 15.02.2021

[88] NIKLAS, C. ; BAUKE, S. ; MÜLLER, F. ; GOLIBRZUCH, K. ; WACKERBARTH, H. ; CTISTIS, G. : Quantitative measurement of combustion gases in harsh environments using NDIR spectroscopy. In: Journal of Sensors and Sensor Systems 8 (2019), No. 1, pp. 123–132. – DOI 10.5194/jsss–8–123–2019

[89] NISHIYAMA, A. ; LE, M. K. ; FURUI, T. ; IKEDA, Y. : The Relationship between In-Cylinder Flow-Field near Spark Plug Areas, the Spark Behavior, and the Combustion Performance inside an Optical S.I. Engine. In: Applied Sciences 9 (2019), No. 8. – DOI 10.3390/app9081545

[90] PASCHEN, F. : Ueber die zum Funkenübergang in Luft, Wasserstoff und Kohlensäure bei verschiedenen Drucken erforderliche Potentialdifferenz. In: Annalen der Physik 273 (1889), No. 5, pp. 69–96. – DOI 10.1002/andp.18892730505. – ISSN 00033804

[91] PIRKER, G. (eds.) ; WIMMER, A. (eds.) ; MEYER, G. (eds.) ; KIESLING, C. (eds.) ; NICKL, A. (eds.) ; TILZ, A. (eds.): Diagnostic Methods for Investigating the Ignition Process in Large Gas Engines. Baden-Baden : AVL List GmbH, 2018 (Proceedings 2018 - 13. Internationales AVL Powertrain Diagnostik Symposium)

[92] PISCHINGER, R. ; KLELL, M. ; SAMS, T. : Thermodynamik der Verbrennungskraftmaschine. Wien : Springer, 2009 (Der Fahrzeugantrieb). – ISBN 978–3–211–99277–7

[93] PISCHINGER, S. : Effects of spark plug design parameters on ignition and flame development in an si-engine, Massachusetts Institute of Technology, Dissertation, 1989

[94] PISCHINGER, S. ; GEIGER, J. ; NEFF, W. ; THIEMANN, J. ; BÖWING, R. ; KOSSWAR, H.-J. : Einfluss von Zündung und Zylinderinnenströmungauf die ottomotorische Verbrennung bei hoher Ladungsverdünnung. In: MTZ - Motortechnische Zeitschrift 2002 (2002), No. 5, pp. 388–399. – DOI 10.1007/BF03227360

[95] POOLA, R. B. ; NAGALINGAM, B. ; GOPALAKRISHNAN, K. V.: Performance of Thin-Ceramic-Coated Combustion Chamber with Gasoline and Methanol as Fuels in a Two-Stroke SI Engine. In: SAE Technical Paper 941911 (1994). – DOI 10.4271/941911

[96] PRAGER, M. : Analytische Modellierung des Betriebsverhaltens eines Gasmotors mit neuem Gaszündstrahlverfahren für hohe Leistungsdichte, Technische Universität München, Dissertation, 2009

[97] QIAO, L. : Experimental and Computational Study of Pre-Chamber Turbulent Jet Ignition for Lean-Burn Engines. Chicago, 2019 (Workshop on Ignition for Internal Combustion Engines)

[98] RAGER, J. : Funkenerosion an Zündkerzenelektroden, Universität des Saarlandes, Dissertation, 2006

[99] REDTENBACHER, C. : Analyse und Optimierung von Vorkammerbrennverfahren für Großgasmotoren, Technische Universität Graz, Dissertation, 2012

[100] ROETHLISBERGER, R. P. ; FAVRAT, D. : Investigation of the prechamber geometrical configuration of a natural gas spark ignition engine for cogeneration: part I. Numerical simulation. In: International Journal of Thermal Sciences 42 (2003), No. 3, pp. 223–237. – DOI 10.1016/S1290–0729(02)00023–6. – ISSN 12900729

[101] RYCHTER, T. J. ; SARAGIH, R. ; LEZAŃSKI, T. ; WÓJCICKI, S. : Catalytic activation of a charge in a prechamber of a si lean-burn engine. In: Symposium (International) on Combustion 18 (1981), No. 1, pp. 1815–1824. – DOI 10.1016/S0082–0784(81)80187–7. – ISSN 00820784

[102] SAYAMA, S. ; KINOSHITA, M. ; MANDOKORO, Y. ; FUYUTO, T. : Spark ignition and early flame development of lean mixtures under high-velocity flow conditions: An experimental study. In: International Journal of Engine Research 20 (2017), No. 2, pp. 236–246. – DOI 10.1177/1468087417748517. – ISSN 1468–0874

[103] SCHNEIDER, A. ; LEICK, P. ; HETTINGER, A. ; ROTTENGRUBER, H. : Experimental studies on spark stability in an optical combustion vessel under flowing conditions. In: Liebl J., Beidl C. (eds), Internationaler Motorenkongress 2016, Proceedings, Springer Fachmedien, Wiesbaden. (2016), pp. 327–348. – DOI 10.1007/978–3–658–12918–7_22

[104] SCHWARZ, L. : Methodische Untersuchung und ganzheitliche Potentialbewertung zukünftiger Antriebssysteme zur CO2-Neutralität im Rennsport, Universität Stuttgart, Dissertation, 2019. – DOI 10.1007/978–3–658–28085–7

[105] SENS, M. ; BINDER, E. ; BENZ, A. ; KRÄMER, L. ; BLUMENRÖDER, K. : Pre-Chamber Ignition as a Key Technology for Highly Efficient SI Engines – New Approaches and Operating StrategiesOttomotoren – neue Ansätze und Betriebsstrategien. Wien, 2018 (39. Internationales Wiener Motorensymposium)

[106] SENS M. ; BINDER E. ; REINICKE P. ; RIESS M. ; STAPPENBECK T. ; WÖBKE M.: Pre-Chamber Ignition and Promising Complementary Technologies. Aachen, 2018 (27th Aachen Colloquium)

[107] SERRANO, D. ; ZACCARDI, J.-M. ; MÜLLER, C. ; LIBERT, C. ; HABERMANN, K. : Ultra-Lean Pre-Chamber Gasoline Engine for Future Hybrid Powertrains. In: SAE Int. J. Advances & Curr. Prac. in Mobility 2 (2019), pp. 607–622. – DOI 10.4271/2019–24–0104

[108] SHIRAISHI, T. ; TERAJI, A. ; MORIYOSHI, Y. : The Effects of Ignition Environment and Discharge Waveform Characteristics on Spark Channel Formation and Relationship between the Discharge Parameters and the EGR Combustion Limit. In: SAE Int. J. Engines 9 (2016), No. 1, pp. 171–178. – DOI 10.4271/2015–01–1895

[109] SPICHER, U. ; KOLLMEIER, H.-P. : Detection of Flame Propagation During Knocking Combustion by Optical Fiber Diagnostics. In: SAE Technical Paper 861532 (1986). – DOI 10.4271/861532

[110] STĘPIEŃ, Z. : A new generation of F1 race engines – hybrid power units. In: Combustion Engines 167(4) (2016), pp. 22–37. – DOI 10.19206/CE–2016–403

[111] SUZIKI, T. ; TSUJITA, M. ; MORI, Y. ; SUZUKI, T. : An Observation of Combustion Phenomenon on Heat Insulated Turbo-Charged and Inter-Cooled D.I. Diesel Engines. In: SAE Technical Paper 861187 (1986). – DOI 10.4271/861187

[112] SUZUKI, K. ; UEHARA, K. ; MURASE, E. ; NOGAWA, S. : Study of Ignitability in Strong Flow Field. In: Günther M., Sens M. (eds) Ignition Systems for Gasoline Engines. CISGE 2016. Springer, Cham. (2016), pp. 69–84. – DOI 10.1007/978-3-319-45504-4_4

[113] SYROVATKA, Z. ; TAKATS, M. ; VAVRA, J. : Analysis of Scavenged Pre-Chamber for Light Duty Truck Gas Engine. In: SAE Technical Paper 2017-24-0095 (2017). – DOI 10.4271/2017-24-0095

[114] TANUMA, T. ; SASAKI, K. ; KANEKO, T. ; KAWASAKI, H. : Ignition, Combustion, and Exhaust Emissions of Lean Mixtures in Automotive Spark Ignition Engines. In: SAE Technical Paper 710159 (1971). – DOI 10.4271/710159

[115] TEODOSIO, L. ; BOZZA, F. ; TUFANO, D. ; GIANNATTASIO, P. ; DISTASO, E. ; AMIRANTE, R. : Impact of the laminar flame speed correlation on the results of a quasi-dimensional combustion model for Spark-Ignition engine. In: Energy Procedia 148 (2018), pp. 631–638. – DOI 10.1016/j.egypro.2018.08.151. – ISSN 18766102

[116] TODSEN, U. : Verbrennungsmotoren. München : Hanser Carl, 2012. – ISBN 978-3-446-42846-1

[117] TOEDTER, O. ; DAHMEN, N. ; WAGNER, U. ; SCHEER, D. ; KOCH, T. ; SAUER, J. : reFuels - rethinking Fuels for CO2 neutral mobility. Digital Platform, 2020 (32nd SIA 2020 Powertrain & Energy)

[118] TOULSON, E. ; HUISJEN, A. ; CHEN, X. ; SQUIBB, C. ; ZHU, G. ; SCHOCK, H. ; ATTARD, W. P.: Visualization of Propane and Natural Gas Spark Ignition and Turbulent Jet Ignition Combustion. In: SAE International Journal of Engines 5 (2012), No. 4, pp. 1821–1835. – DOI 10.4271/2012-32-0002. – ISSN 1946-3944

[119] TOULSON, E. ; SCHOCK, H. J. ; ATTARD, W. P.: A Review of Pre-Chamber Initiated Jet Ignition Combustion Systems. In: SAE Technical Paper 2010-01-2263 (2010). – DOI 10.4271/2010-01-2263

[120] TOWNSEND, J. S. E.: Electricity In Gases. SCHOLAR SELECT, 2015. – ISBN 9781296541934

[121] TOYAMA, K. ; YOSHIMITSU, T. ; NISHIYAMA, T. ; SHIMAUCHI, T. ; NAKAGAKI, T. : Heat Insulated Turbocompound Engine. In: SAE Technical Paper 831345 (1983). – DOI 10.4271/831345

[122] TRZESNIOWSKI, M. : Rennwagentechnik: Grundlagen, Konstruktion, Komponenten, Systeme. Wiesbaden : Springer Vieweg, 2014 (ATZ/MTZ-Fachbuch). – ISBN 978-3-658-04919-5

[123] TURKISH, M. C.: 3 - Valve Stratified Charge Engines: Evolvement, Analysis and Progression. In: SAE Technical Paper 741163 (1974). – DOI 10.4271/741163

[124] WANG, J.-P. ; CHO, W. D.: Oxidation Behavior of Pure Copper in Oxygen and/or Water Vapor at Intermediate Temperature. In: ISIJ International 49 (2009), No. 12, pp. 1926–1931. – DOI 10.2355/isijinternational.49.1926. – ISSN 0915-1559

[125] WEBERBAUER, F. ; RAUSCHER, M. ; KULZER, A. ; KNOPF, M. ; BARGENDE, M. : Allgemein gültige Verlustteilung für neue Brennverfahren. In: MTZ - Motortechnische Zeitschrift 66 (2005), No. 2, pp. 120–124. – DOI 10.1007/BF03227253

[126] WEISSNER, M. ; BERGER, F. ; SCHÜTTENHELM, M. ; TALLU, G. : Lean-burn CNG engine with ignition chamber: from the idea to a running engine. In: Combustion Engines 176(1) (2019), pp. 3–9. – DOI 10.19206/CE–2019–101

[127] WESTERHOFF, M. ; SPRINGER PROFESSIONAL ENGINE TECHNOLOGY (eds.): Formula 1 Engine from Mercedes with over 50 Percent Efficiency. https://www.springerprofessional.de/engine-technology/race-cars/formula-1-engine-from-mercedes-with-over-50-percent-efficiency/15061334. Version: 15.09.2017, Access Date: 31.01.2021

[128] WIPPERMANN, N. ; THIELE, O. ; TOEDTER, O. ; KOCH, T. : Measurement of the air-to-fuel ratio inside a passive pre-chamber of a fired spark-ignition engine. In: Automotive and Engine Technology 5 (2020), No. 3-4, pp. 147–157. – DOI 10.1007/s41104–020–00067–w. – ISSN 2365–5127

[129] WIPPERMANN, N. ; TOEDTER, O. ; KOCH, T. : Optical Measurement of Spark Deflection Inside a Pre-chamber for Spark-Ignition Engines. In: SAE Technical Paper 2020-01-5096 (2020). – DOI 10.4271/2020–01–5096

[130] WOHLGEMUTH, S. : CO2-optimierter Antrieb eines Kleinfahrzeuges, Technische Universität München, Dissertation, 2016

[131] WOSCHNI, G. ; SPINDLER, W. ; KOLESA, K. : Heat Insulation of Combustion Chamber Walls — A Measure to Decrease the Fuel Consumption of I.C. Engines? In: SAE Technical Paper 870339 (1987). – DOI 10.4271/870339

[132] YU, S. ; XIE, K. ; YU, X. ; HAN, X. ; LI, L. ; LIU, M. ; TJONG, J. ; ZHENG, M. : The Effect of High-Power Capacitive Spark Discharge on the Ignition and Flame Propagation in a Lean and Diluted Cylinder Charge. In: SAE Technical Paper 2016-01-0707 (2016). – DOI 10.4271/2016–01–0707